D0342269

THE
TIDES
OF
MIND

THE
TIDES
OF
MIND

Uncovering the Spectrum of Consciousness

David Gelernter

LIVERIGHT PUBLISHING CORPORATION

A Division of W. W. Norton & Company

Independent Publishers Since 1923

New York · London

For information about permission to reproduce selections from this book,
write to Permissions, Liveright Publishing Corporation,
a division of W. W. Norton & Company, Inc.,
500 Fifth Avenue, New York, NY 10110

For information about special discounts for bulk purchases, please contact
W. W. Norton Special Sales at specialsales@wwnorton.com or 800-233-4830

Manufacturing by RR Donnelley Harrisonburg
Book design by Chris Welch
Production manager: Louise Mattarelliano

Library of Congress Cataloging-in-Publication Data

Names: Gelernter, David Hillel, author.
Title: The tides of mind : uncovering the spectrum of
consciousness / David Gelernter.
Description: First Edition. | New York : Liveright Publishing
Corporation, 2016. | Includes bibliographical references and index.
Identifiers: LCCN 2015037465 | ISBN 9780871403803 (hardcover)
Subjects: LCSH: Consciousness. | Cognition.
Classification: LCC BF311 .G396 2016 |
DDC 128/.2—dc23 LC record available at http://lccn.loc.gov/2015037465

Liveright Publishing Corporation
500 Fifth Avenue, New York, N.Y. 10110
www.wwnorton.com

W. W. Norton & Company Ltd.
Castle House, 75/76 Wells Street, London W1T 3QT

1 2 3 4 5 6 7 8 9 0

For my Jane
and in loving memory of my father,
Herbert Gelernter, zichrono li'vracha.
One of the six men who invented AI,
and that was the least of it.
I shall not look upon his like again.

Contents

Preface

I have practiced computer science for thirty years. What drew me to the field was the unlimited plastic power of digital computers: computers give you the power to dream up almost any machine you like, shape a simple version in modeling clay, and then flip a switch and watch it come alive. This naïve-sounding vision is almost real, almost true. A good programmer can sit down at the keyboard and build a program—a working piece of software—with nearly the complexity of an aircraft carrier *all by himself*, to his own designs and no one else's. The fact that you can achieve so much *all alone* is one good reason to be fascinated and terrified by computing. The field has always attracted sociopaths.

But there are also good scientific reasons to be intrigued by this enormous power. For example, how does one study the mind?

Here is a strange fact about mind study. It demands explanation, although people in the field take it for granted and barely notice its strangeness.

Go to a well-stocked academic library and grab, at random, some books and journals on philosophy of mind. Virtually *everything* you pick up will be full of digital computers, full of assertions about the science and technology of computers and software, and results from the theory of computing.

Why should philosophy of mind be obsessed with digital computers? Philosophical ethics is not obsessed with pneumatic jackhammers. Political philosophy is not obsessed with fiberglass fishing rods. Why should philosophers of mind return to a certain machine and its capacities again and again?

There are three explanations, all related. One centers on computers as a test-bed for mind theories. Another focuses on computing as a powerful, simple way to describe or blueprint events in time: *processes*—that is, organized actions. The last explanation, a theory called computationalism, asserts that brains *are* computers, and the mind is just software that runs on the brain. This would be awfully neat if it turned out to be true.

Now, if you had decided that a jackhammer happened to be a precise model of an individual driven to the breaking point by an insoluble ethical dilemma, or a fiberglass fishing rod was the ideal way to understand an electorate weighed down with misinformation but still endowed with the springiness of freedom, these technologies might be important to their respective areas of philosophy.

Regarding philosophy of mind, serious thinkers have concluded that certain machines bear a remarkable resemblance to the human brain: computers resemble brains, and minds are like software. The alleged resemblances are no coincidence, of course. Scientists designed these machines to carry out tasks for which, *ordinarily*, they used their minds. So it was perfectly natural to imagine that when *machines* carried out these tasks, those machines were showing themselves to resemble mind or brains.

Let's briefly look at the first two reasons for attending to computers and their capabilities. (They are my own reasons for going into the field of computing.) The third requires more space but is fundamental to the whole drift of modern thought.

Mind researchers saw, in computing, an opportunity to test their mind theories directly—to embody them in working models, throw the "on" switch and see what happened. Did the working models perform as predicted? For example, by carrying out a particular theory embodied in a *working program*, did the computer succeed in learning grammar and sentence structure from the spoken language all around it? Or did it succeed in planning a series of movements to accomplish a goal, or create memories of the expected type, or learn to pronounce words it had never heard before? Did your theory *work*?

This kind of programming is one branch of "artificial intelligence," or AI. This is "theoretical AI," which centers on the human mind. "Applied AI" seeks to solve problems that minds can solve only by using intelligence, *not* by following a prearranged set of rules. Sometimes applied AI works by incorporating mind-like techniques into software.

Applied AI has scored major successes. Two of its most important in recent years were the work of IBM Research: the "Deep Blue" program that won the world chess championship in 1997, and the "Watson" program that beat the best *Jeopardy!* players in the world in 2011. (IBM has a long history of pathbreaking work in AI, going back to its famous geometry theorem prover of 1957—the third AI program ever built, and the first that *did* anything.) Many other enormously impressive accomplishments have come out of applied AI, from the sophisticated robots that are becoming ubiquitous, to software that can invent syntheses for organic compounds (can figure out, that is, how to produce complex chemicals from raw materials), to other programs that do jobs ordinarily requiring a scientist in the flesh, PhD in hand.

The second reason for computing's central place in modern research is that it provides the framework for understanding *processes*, actions-in-time. The root idea is nearly as simple as ideas get. Given a list of steps, carry them out in sequence. Finish one and then proceed to the next, until you reach the end of the list. Then stop.

Two simple details increase the power and expressiveness of this idea enormously. One is *recursive structure*. Given a list of instructions, any individual instruction can be replaced by an entire list. (This is like saying that any number or variable in an algebraic expression can be replaced by a whole expression: I can start with $x + y + 5$ and replace the y so that I have, say, $x + (3x + 120 + z) + 5$, and so forth. I still have a legitimate expression.) And I can vary the meaning of the list of instructions by setting parameters. If I set x equal to 150 and y to 14, then $x + y + 5$ means (or "equals") 169. If I set x and y equal to 2 and 3, respectively, then the same expression equals 10. Changing the values of parameters (in this case the parameters x and y) allows one expression to mean many things. Lists of computing instructions work the same way. The result is a powerful and concise way of capturing actions-in-time.

These were my own reasons for studying computation: the modeling power of software, the idea of algorithm.

Like so many others, I have always been obsessed with the mind. When I was a student in the 1960s and '70s, the whole world spoke Freud's language of repression and childhood trauma, the unconscious and the egomaniac, phobias and libidos and Oedipus complexes—just as the whole world does today, only we no longer bother to mention Freud.

As an undergraduate I studied neurophysiology and molecular biology, the usual things. But those left me a long way from the mind, barely able to make it out in the distance. The *brain* doesn't believe things, or get excited or grow wistful or daydream about a farm by the ocean in Maine, with a studio overlooking the sea. The *mind* does those things. John Milton wrote, "The mind is its own place." The mind is the landscape we invent, the landscape of *us*.

AI attracted me to computer science because (again, like so many others) I had a theory I was eager to test. The theory led to a software project when I was a young professor at Yale—carried out mainly with Scott Fertig, one of my graduate students in the late 1980s and '90s. The project was successful—but not in the way I had hoped. It became a suggestive and promising application, one of the first that converted data records (describing chest X-rays, in our case) into online advice about new cases, advice based on general rules derived from the input data and backed up by concrete examples, also from the input data.

But I had wanted to build a program with a dial in front marked "focus." You could vary the value of focus by turning the dial, from maximum to zero. At maximum focus, the program would "think" rationally, formally, reasonably. As focus diminished, its "mind" would start to wander—it would attend to other things than the patient-case right before it. As you kept dialing focus lower and lower, the program's mind wandering would grow more pronounced. Finally, it would start to free-associate—and finish by ignoring the user completely as it cruised off into its own mental adventures.

The actual program did some intriguing things. But it never achieved the sort of focus-knob behavior I had hoped for. Mainly, it demonstrated that I didn't yet understand my own theory.

So, back to the drawing board. Nearly twenty years later, this book is (at last) the result.

An author must review other, competing ways of viewing his topic. I will do that partly in this preface, partly in the body of the book. But in my case, there *are* no competing views to my theory of mind. By which I mean, there are no others I must reject if my view is to be successful. The reason is simple. The others are arguing an important question, focused (let's say—metaphorically) on *what is the best route into the city from the north?* But I am coming from the east.

I will argue that both questions (how to come from the north *and* the east) are important: we need *both* views of the mind. In a way, the two approaches are orthogonal: independent ways of examining the same topic—in principle, complementary.

Before proceeding to reason three for mind studies to be obsessed with computation, one general point. A computer scientist working on mind must discuss philosophy; a philosopher of mind must discuss computation. I was trained as a computer scientist. But intellectual mingling and fraternization across this line has been the way of the world since the birth of AI.

This blurring of lines between computation and philosophy of mind makes perfect sense. In the short history of artificial intelligence, there has always been much mixing along the boundary between philosophy of mind and AI. After all, it was a computing expert and mathematician—Alan Turing, the genius who invented artificial intelligence (and many other things) and introduced it in a philosophy journal called *Mind*—who did more than anyone to nudge philosophers in the direction of digital computers.

In general, academics say they love this sort of cross-field fertilization. In fact, they hate it. (Not *all* of them, obviously. Many of them.) There's nothing surprising in that. An academic is nothing if not a specialist.

Finally, to the third reason for the intimacy between computing and philosophy of mind: a tremendously popular and influential theory of the mind called "computationalism." Computationalism is the intellectual project that opened the floodgates, that brought ideas and language from computing and software roaring, pounding, and exploding (in bright plumes of spray and the odd leaping fish) into the peaceful green fields of philosophy.

Computationalism asserts that computing is the very stuff of mind: that the brain is a sort of organic computer, and the mind is like software that runs on the brain.

Some people and many computationalists believe that you can build a mind out of software. A real, *genuine* mind. If you arrange enough computer instructions correctly, the app or the program you have just built will be a mind. If it is running on your laptop, your laptop *now has a mind*. Those who believe this, many of whom are computationalists, mean a *complete* mind—one that can think and solve problems of all sorts, but can also *feel*, can experience the world, can *imagine*. If you told your mind-equipped computer to "picture a swan," it would picture a swan just as you would. In fact, your computer would be *conscious*, just as you are.

Why would anyone believe this? We need to go further: Why would this be the *consensus view* of the *intellectual mainstream in the mind sciences*? Answering this question will help put this book in perspective—in the context of today's mainstream views of the mind.

The ideas of computing seem essential to the study of mind because of a state of affairs at the very start of the field in the late 1940s and early 1950s. Computers were invented to solve problems that would help win the war. They solved those problems and *did* help win the war.

As researchers emerged from wartime and began to think in broader terms about the science and technology they had discovered, something hit them. What did it mean for a person to think? It meant (many decided) to *compute*, broadly speaking, using a large range of different methods. (To compute meaning simply to *calculate*; no suggestion, yet, of digital computing machinery.)

Thinking—*rational* thought, or *reasoning*—meant *computing*, and computing meant following some sort of rule, or making one up. Maybe you are solving a high school algebra problem. Maybe you are planning your day, or figuring out where you left your keys, or how to clear a tree out of the driveway knocked over by last night's storm. The essence of *rational thought* is building your case step by step so that each step is justified by the previous ones. The variety of techniques or rules to choose from is huge, but there's always some kind of rule.

Sensational news! (Stop press!) Because now, for the first time in history, people were not the *only* ones that could follow rules.

Digital computers could too. This was exactly what digital computers were for.

Accordingly, it struck researchers—not everyone in the still-minuscule field, but more than a few—that computers were not merely programmable digital calculators. Not merely calculators whose behavior could change. Computers were not merely calculating machines of any kind. They were *thinking* machines. Why? Because thinking—rational, reasonable thinking—was really just computing. It all came down to *computing*. And digital computers could do any computation there was.

Everyone knew that a human mind could do more than rational thinking. But rational thinking seemed to be its defining activity. The word that especially fascinated Turing was "intelligence." Rational thinking was the manifestation of intelligence. Turing knew well that there was more to a mind than intelligence. But intelligence was the main thing. That's why the field has the name it does: not artificial mind, artificial thought, artificial reasoning, but artificial *intelligence*.

For Turing, and a few other scientists in Europe (mainly Britain) and America, the general shape of the artificial intelligence project was clear. First you got computers to manifest intelligence in many areas—not just mathematics, but (for example) in articulate, wide-ranging conversations on any topic. Then you could fill in the rest of the mind, to the extent filling was needed: emotions, sensations, attitudes, many other mental states, even consciousness.

Since mind is for *rational thought*, which amounts to computing, which can be accomplished by computers, to study the theory and structure of digital computers and software *was* to study the essence of mind. Before long, the philosophical field of computationalism emerged. Computationalism involved more than these simple intuitions. But these intuitions were the heart of it.

Nearly all computationalists believe that *minds relate to brains* as *software relates to computers.* This analogy was crucial because one of the hardest of all points in philosophy of mind had been just this: How do minds relate to brains? How can a mere *thought* (I think I'll type the letter *R*)—intangible, immeasurable—be converted into physical action in the real world? How are casual, passing fancies converted into physical motion, into the complex nerve signaling and muscle movement that is typing? What could the connecting link look like? What could it *be*? The analogy offered a kind of answer: mind relates to brain as software relates to computer.

Computationalists say this: to understand the mind, study software. A digital computer is merely a set of binary switches wired together in complex ways. A binary switch is just like an ordinary light switch. The switch on your wall is binary, with two positions: "on" and "off." Whatever position we put it in, it remembers. Computers and computer memories are built out of microsized, purely electronic (no moving parts) versions of a light switch.

The active, thinking part of the brain is built out of neurons, and they can be described as binary switches too. A neuron is either off or on. On, it transmits a nerve signal to all the neighboring, downstream neurons. Off, it transmits nothing. Neurons turn on when the right signals reach them from their neighbors upstream, some ons and some offs.

With a little imagination, then, a brain becomes a kind of organic computer; and the mind is the software of the brain.

Concretely, building the right software, and downloading it on the right digital computer, will yield a computer with a conscious mind that is just as capable as the human mind. There will be differences, but *basically* the two types of mind can do the same things. All digital computers are identical except for performance, meaning speed and memory size, and details that are irrelevant here, such as power consumption. We might need a very fast computer to produce a mind, but that superfast computer will be merely a speeded-up version of our computers today. There will be no logical difference. And the software that *creates* our computer mind will be built of the same parts as today's software. Build the software right (say computationalists), run it on the right computer, and you will have a mind. Not a simulated mind, not something *like* a mind, but a real, working mind.

Dissenters say this: We only need to build the right software to have a mind? And run it on a fast enough computer? But this mind-creating software would be built of the same parts, the same basic *instructions*, as today's software. And we know what those instructions do. They move numbers around ("Move a number from *here* in memory to *there* in memory"), do simple arithmetic ("Add these two numbers"), and logical tests ("If the result is zero, skip the next ten instructions and continue"). Any person can do all these instructions—much more slowly, but just as correctly, as a digital computer.

So let's do a thought experiment, say the dissenters—let's imagine a simple test. Someone hands you the remarkable software application that creates minds on digital computers. You can read each instruction and carry it out just as a computer can. So you sit down at a table with paper and pencils and carry out the first five instructions. (Maybe you add some numbers and multiply some others.) Has a new mind been created as a result of those five instructions? Of course not. You can't create a mind by adding a few numbers. Then you do the next ten. Then (in a burst of enthusiasm) the next two hundred. You've filled up a whole pad of paper and resharpened your pencil twice. But have you created a mind? No. If ten additions, multiplications, logical tests, and so forth don't create a mind, why should two hundred? You can proceed to another three hundred instructions, and another and another. You just keep adding numbers, moving them around, and so forth as the software tells you to. When does a new mind get created? Does it pop into being after the ten millionth instruction? The ten trillionth? No.

How could it? How could it *possibly*? How could playing with num-bers (just as a digital computer does) *ever* produce a mind? You can imagine the entire process, say the dissenters; you can imagine it in exact detail, because each instruction is precisely defined and simple enough for a child to do. And, by imagining the whole process, say the dissenters, you can see that *no new mind is produced, ever.* We are doing nothing that could *possibly* create a mind.

The computationalists have a proposition, say the dissenters, like winning the marathon at the Olympics by hopping up and down and croaking like a frog. If you hop a hundred times, will you win? No. If you hop a thousand times, or a million? No. *No.* Why not? Because hopping up and down and croaking like a frog has nothing to do with winning a marathon at the Olympics.

The computationalists' answer is this: Imagine a single neuron. (You can't see it with the naked eye, but it's there.) Can *it* think? Understand? Create consciousness? Of course not. Can a hundred neurons? A thousand? A million? No! The idea seems ridiculous. Yet we happen to know that when you have *enough* neurons, a hundred billion or so, and they're connected in the right way and attached to a body—those neurons do, indeed, create consciousness. So the inability of one, a hundred, a million, or a hundred million com-puter instructions to create consciousness is completely irrelevant.

To which the dissenters reply, "So what?" True, a hundred bil-lion neurons, connected correctly and wired to a body, will yield consciousness. But that doesn't mean that a hundred billion random *anythings* will create consciousness! Neurons work, but why should I believe that computer instructions (of all things) will do the same job? Why shouldn't I believe the thought-experiment evidence that tells me they don't? After all, a hundred billion grains of sand don't create consciousness, or a hundred billion mosquitoes, or sardines, or flamingoes, or anything whatsoever—*except neurons.*

That's where the argument stands today. Whether you are a computationalist or a dissenter, the ideas of computing have merged with those of mind science. The approach I take in this book is radically different from computationalism, and far away from these arguments. But readers ought to know where things stand in the world at large.

The computationalist view has too many leaders and persuasive exponents to list; I refer to some in the course of the book. The two leaders of the dissenting camp are John Searle and Thomas Nagel—but their views are very different.

In 1980, Searle published a famous thought experiment called the Chinese Room,[1] which is similar, in essence, to the thought experiment I have described here. Searle's argument was fallen upon immediately, attacked from every side, like Caesar in the Capitol—but with fury more than considered passion. Searle's particular focus is understanding. No computer (he believes) will ever understand anything at all, no matter what it *seems* to do. Computer instructions just don't have it in them, he believes, to create understanding—and (as I've said) he uses a thought experiment like the one I have used (his came first!) to make that point. He is a thoroughgoing materialist; he has no interest in metaphysical or spiritual claims. He only insists on the prudent skepticism that has always been fundamental to science.

Thomas Nagel takes a broader, in some ways higher-level, view.[2] His argument is too wide-ranging to summarize here. But he does not believe that computers are capable of creating subjectivity: your own personal experience, your mental life, your own private landscape of mind—the world inside your head that no one but you can ever wander through, ever see or come to know or directly experience in any way at all. He does not believe that computers are capable of creating consciousness. (Searle is with him so far, although Nagel's emphasis on subjectivity is different from Searle's on understanding and the mental property called "intentionality" or "aboutness"—a belief is *about* something.)

But Nagel believes, further, that a scientific revolution will probably be required before we have the means to explain consciousness. Consciousness and other aspects of the mind, he notes, raise hard questions about the whole smooth, shiny Spandex cover that science has stretched over the bumpy reality of nature. Nagel, like Searle, is a strict materialist, a deep believer in science as far as it goes. He rejects metaphysical, spiritual, or religious explanations of the universe. But he believes we are nowhere near a convincing explanation of subjectivity or consciousness. We can't even say what a convincing explanation would *look* like.

A last significant dissenter, not as influential as Searle and Nagel but important in his own right—Colin McGinn—believes, like Nagel, that science as it stands lacks the ideas and the intellectual framework to explain subjectivity and consciousness.[3] Unlike Nagel, McGinn believes our problem lies deeper. Sheep have never understood Gilbert and Sullivan, and parrots (who are thoughtful and brilliant) do not and *cannot* understand physics, or how to win at chess, or how to read. They can try as hard as they like (and parrots try very hard), but their brains are just not cut out for it. *Our* brains have limits too, says McGinn. Not only do we not understand consciousness; there are no grounds for believing we ever will.

I will discuss the phenomenologists and the Freudians in the book itself; their views are closer to mine. The Freudians have kept the serious study of the human mind alive, have kept "depth psychology" alive (as a non-Freudian psychiatrist described Freudianism during Freud's own career) in an era that often seems contemptuous of the individual and what sets him apart from the crowd.

When I was a child, I loved science but found it discouraging: it seemed as if all the questions already had answers. Whatever I asked my parents or their scientist friends or my teachers, there always seemed to be answers—from why is the sky blue to what are stars made of, and why is metal shiny, and how fast can elephants run, and can pigs stand on their snouts?

Learning about science and technology, to the small extent I *have* learned, has been equally a matter of learning what we know and what we don't. It is hard for a child to accept (and hard for adults too) that deeply important facts about human beings and our world lie right in front of us, hidden in plain sight. But we know that blasé boredom and sophisticated cynicism are the easiest and cheapest of human attitudes, surprise and wonder the most precious. Not for nothing do we remember the luminous visions of childhood as irreplaceable, different in kind from the rational (or fairly rational) worldview of adulthood, and our best compensation—payment in advance—for whatever is to come.

On the other hand, rationality is not so bad either.

TRANSLATIONS FROM FRENCH and German literary texts are mine, except for Kafka and for Proust's *À la recherche du temps perdu*, where I have followed the lead of the Moncrieff, Kilmartin, and Mayor translation (*Remembrance of Things Past*, Random House, 1981). Translators from the Russian and a few other instances are given in the list of cited literary works at the end.

THE
TIDES
OF
MIND

One

The Tides of Mind

M any thinkers see the mind as a massive ancient temple newly unearthed in a desert somewhere. Dozens of world-class teams are spread all over the site in shorts and floppy hats, each team absorbed in its own courtyard or cellar or fortified guard post, measuring, remeasuring, taking photos, measuring again. Someday, one group will issue the definitive work on all the courtyards or all the cellars or all the guard posts, and eventually the whole story will emerge—the truth—and the mystery of mind will be solved.

But this is all wrong. The mind *changes constantly* on a regular, predictable basis. You can't even see its developing shape unless you look down from far overhead. You must know, to start, the overall *shape* of what you deal with in space and time, its architecture and its patterns of change. The important features all change together. The role of emotion in thought, our use of memory, the nature of understanding, the quality of consciousness—all change continuously throughout the day, as we sweep down a spectrum that is crucial to nearly everything about the mind and thought and consciousness.

This book is about that spectrum and a new way to understand the mind. This new way incorporates the findings and observations of many thinkers but rests ultimately on the solid, unspectacular bedrock of common sense.

If you understand the nighttime sky, you understand how the stars' positions *change*. Not to understand those patterns of change is not to understand the sky. The mind changes too.

Even a child knows that the mind at work on a math problem on a fine sunny morning is a strangely, strikingly different instrument from the same mind fighting through a nightmare, in the grip of terrifying hallucinations and oblivious of the outer world. We know that our thought processes differ when we are fresh and wide-awake, when we are at a comfortable midday midpoint, and when we are drifting off to sleep. We *all* know this. But, like an island completely submerged and then released by the ocean every day, we go from one state to the other in a series of indiscernibly gradual, gently lapping steps.

The very purpose of the mind changes as we move down-spectrum, from *doing* to *being*, from the mental manipulations called "thought" to sensations and feelings that reach, at the absolute bottom, the pitch of pure un-self-conscious being, of wholly unreflective experience. As we descend from the top, our gift for abstraction and reasoning fades while sensation and emotion begin to bloom cautiously and then grow lusher and brighter. (Up-spectrum, we keep emotion on a short leash because it disrupts thought.) As focus falls and conscious mind relaxes its guard, and memory ranges more freely, the mind wanders. Daydreams grow more insistent. And we *experience* more of the saturated intensity of emotions and sensations—barefoot in the clover and wildflowers. Consequently, we *reflect* less and less, our self-awareness declines, we lay down fewer and shoddier memories. Our minds are dominated by sense and feeling.

Where the spectrum bottoms out, we find dreams. They are "intensely emotional," writes the neurophysiologist J. Allan Hobson.[1] Of course, they are hard to recall; sometimes they erase themselves *while* we try to remember them. ("That night I was startled

Spectrum	High focus	**MEMORY** use is disciplined, focused	**THOUGHT** is rational **REFLECTION, SELF-AWARENESS** are strong	
	Medium focus	**MEMORY** use ranges freely, wanders occasionally	**THOUGHT** seeks experience	*emotions emerging* *daydreams beginning*
	Low focus	**MEMORY** takes off on its own	**THOUGHT drifts REFLECTION, SELF-AWARENESS** are weak	*emotions blooming* *dreaming*

FIG. A. *A quick sketch.*

awake by a dream," writes the psychoanalyst Stephen Grosz, "which began to dissolve as soon as I woke.")[2] In fact, we remember the intensely vivid, emotional experience of earliest childhood badly too. Dreams at their most saturated make time run backward. The effect can be overwhelming. But up-spectrum, where creating memories is the main business of mental life, we can almost hear each event we enact tolling endlessly downward through the future as recollections reenter consciousness over and over and over.

These transformations govern daily life. As children mature, they seem to push their way up a broadly similar spectrum. The same idea can even be applied to societies as a whole. They have their favorite thought styles—old ways to think versus fashionable new ways. Thus T. S. Eliot's striking comment on Dante: "Dante's is a *visual* imagination. It is visual in the sense that he lived in an age in which men still saw visions" ("Dante").

Ancient literature drifts farther and farther out of focus to modern minds—not just because old literature is written in old language, not just (by no means!) because it uses unfamiliar assumptions about society and each person's status and value, but also, most important by far, because it uses different thought styles from those of today. We favor high-focus, analytic thought. Our elite thinkers cluster around the top of the spectrum, whether or not they belong there. (That's another story.) Older societies favored lower-spectrum approaches.

Trying to read ancient literature (such as the older strata of the Hebrew Bible) without retuning our minds to lower-spectrum settings is a plain mistake—like listening to an old recording at the wrong speed, or watching a movie at the wrong frame rate, in the wrong aspect ratio, that's been dubbed. The brain doesn't change over the centuries, but the mind *does*, subtly, as habits of thought and the qualities of consciousness we cultivate change.

The spectrum, however, can help us grasp what has changed over the generations. It can help us see why being a small child is so important to our private mental histories: the world dazzling bright and deepest dark all around us, full of mystery. "Those fleeting moods of shadowy exaltation," writes William Wordsworth, caught up in childhood memories. "Oh! Then, we feel, we feel" (*Prelude*, Book 5). The spectrum is the first thing we need to know about the quality of consciousness. Arguably, it is the first thing we need to know about the mind in general.

Yet mind scientists and mind philosophers ignore the spectrum almost completely. They ask good questions, but they *ignore* good questions too. What are the mind's dynamics? How do relations between thinking and memory change over a day? How does the role of memory itself change, between its duty as mainly an information source up-spectrum (where did I put it, what do I do next, who is that?) and its chattier, storyteller role down-spectrum, supplying remembered incidents, anecdotes, and eventually the whole rich ambience of dreams? Every day we cross the hallucination line as we approach sleep. Past that line, our ideas and recollections become real as they occur to us, in the very process of our conceiving them. You can see it happen if you watch closely. How will we understand this portentous transition if our view of the mind is static? And if we refuse to study mind dynamics over the course of just one day, how will they help us understand the changes we experience as we grow up? Or over a lifetime?

These questions have rarely been asked, even though the answers are so simple, so important, and—in bare outline—so obvious! They stare us in the face. Every day we descend a continuous spectrum of mental states like shimmying down a rope—a spectrum of qualities of consciousness, from the wide-awake, highly focused state in which we do our best reasoning and analysis, through increasingly diffuse states in which our thought processes (or trains of thought, thought sequences) are more likely to be interrupted, and finally into the free-flowing, free-associative thinking that leads straight into hallucination, sleep, and dreaming. I am speaking here about the conscious mind, but the functions of the unconscious mind are mapped out by the spectrum too. Blocking out the conscious mind's functions will let us see what the unconscious mind is doing.

What's Hard about This?

The facts are simple and obvious, and they haven't been missed because of obtuseness. So how *have* they been missed? How—for that matter—could there *be* a new way to understand the mind? How could philosophy, science, and plain curiosity have missed anything during all those long centuries of ransacking the merchandise since Descartes, since Plato, since the dawn of man? How could any bargains be left?

The problem is subjectivity. The problem is our strange position *inside* the phenomenon that we are trying to understand. It is hard to track the rising tide when you are in the water.

One more reason for overlooking the obvious facts is so simple that we are likely to miss it. As we descend the spectrum into the circus din of vivid, sometimes bizarre hallucination, our attention grows overstrained, sensation and emotion fill our mind to the edges—and we are less and less able to create sound new memories. We don't pay as much attention as we should to the lower spectrum, because it's so hard to remember what happens there.

Room with a View

The mind is a room with a view: from inside, we observe the external world *and* our own private, inner worlds. Mentally, we are stuck inside our rooms as we are stuck, physically, within our bodies. The view is great—and had better be, because we can't ever leave.

"O that you could turn your eyes to the napes of your necks and make but an interior survey of your good selves!" says the voluble arch-politician Menenius in *Coriolanus*, Shakespeare's masterpiece of steel-gray heat. "O that you could!"

Many of the largest and deepest questions in philosophy center on this two-part reality of the mind: a *room* with a *view*. The glittering winter's dawn outside versus the warmth and light within. Kant builds one version of his two basic, eternally true "intuitions" on *inner* mind versus *outer* reality: the idea of space underlies our intuitions of the external world. Yet even before space, there is time, which makes the inner world comprehensible.

In recent years, however, mind researchers and philosophers have tended to downplay or just ignore the room in favor of the view. Pure objectivity is their holy grail, and subjectivity seems suspiciously good friends with (almost) the worst character in the whole world, the *unscientific*. "The history of philosophy of mind over the past one hundred years," writes the philosopher John Searle, "has been in large part an attempt to get rid of the mental by showing that no mental phenomena exist over and above physical phenomena."[3] This focus on the physical over the mental seems supremely scientific and has an inevitable consequence. As Searle writes elsewhere: "This crisis produces a flight from subjectivity."[4] A flight from subjectivity: we ignore the room and care only for the view.

Computer-based ideas of mind have encouraged us to disregard and cold-shoulder subjectivity, and computer-based ideas continue to dominate the field. Yet subjectivism has more defenders than it did a generation ago. Important voices like John Searle's and Thomas Nagel's continue to insist from inside the philosophical mainstream (in very different ways) on the importance of subjective reality. Today, more people are listening. (Not many more, but any progress is welcome.) From outside, subjectivism is championed by phenomenology—a school of the early 1900s, now reviving.

Still more important is the philosopher and psychoanalyst Jonathan Lear's reading of Sigmund Freud, whom he calls the inventor of the "science of subjectivity."[5] Freudianism is staging a strong though quiet comeback, on the basis of a few simple, core ideas that not even the most florid anti-Freudian can deny with a straight face. But depth psychology (Freud's invention), and the whole field of subjectivism in science and philosophy, have been marking time for decades, playing defense. "Tidal psychology," "spectrum psychology," "daily mind tracking" (take your pick) has yet to be born.

Freud might have invented the science of the subjective; obviously, he did not invent subjectivity. Mental life is subjective by definition. Private experience can *only* be subjective. And the mind creates private experience. So the science of mind must be a *subjective* science. We want neurobiology to explain the phenomena we've discovered, but first *we must discover them*, and be sure of them.

How Subjective *Is* Subjective?

If we are serious, we can't take anything for granted. Just how subjective *is* subjective?

Often other people know just what we think and feel—because we tell them. Sometimes we do it on purpose; other times, implicitly—in words and with our faces and bodies. "The human body," writes Ludwig Wittgenstein, "is the best picture of the human soul."[6]

Sometimes other people know what we feel better than we do. Jack is a middle-aged man I know who takes a battery of medications for chronic pain. None relieves the pain absolutely, and the medications take hold and wear off gradually. On certain occasions his wife will ask, "Are you sure you took your meds this evening?" "Of course I did; I feel fine!" Jack will snarl. Then he will march back into the bedroom to establish that she is wrong—and discover, usually, that she is right. The pills will be laid out on the pill shelf, untaken. His wife knows his pain level better than he does.

We know intellectually how other people feel. Still more important, we *feel* each other's feelings; we sympathize—we "feel *with*." The reason we can feel other people's feeling is that we are feeling creatures ourselves, and we know how *we* feel when we say certain things, look certain ways. In a sort of emotional resonance, we can—under the right circumstances—feel someone else's feelings in our own bodies.

It is a philosophical conundrum that what you call red, I might experience as blue, while I see "blue" as red. Our subjective experiences of color might be radically different, and neither of us would ever know.

Yet I can see that you smile slightly, and frown ironically, and sneer thoughtfully, in roughly the same circumstances I would—at least if you and I are close and understand each other. We describe ourselves constantly, and we try hard to be understood. Colorful clichés—butterflies in the stomach, insides twisted in knots, jumping for joy, bored to tears, bursting with news—help us to be understood. "My heart aches, and a drowsy numbness pains / My sense . . ." (John Keats, "Ode to a Nightingale"). Mental life *is* irreducibly subjective, but we know plenty about each other's mental states. Feelings *can* arc gaps. Feelings do flow, sometimes, from one body to another.

Of course, our knowledge of other people can't, ultimately, go further than they allow, and it can never go all the way. We all know secrets about ourselves that we have never told and never will. Usually we die, I believe, with our deepest truths unspoken.

> "We are most of us," I said, "in some kind of agony." (Martin Amis, *London Fields*)

We must operate the best we can with the knowledge we can get—in the world that *is*. We have more than enough knowledge to go much further than we have in understanding the subjective world of the spectrum. We must start our study by knowing what the mind is like from inside.

Mind from Inside

What I have said about the centrality of subjective experience amounts to a type of phenomenology, the philosophy developed by the Czech

Jew Edmund Husserl (1859–1938). It took off in several different directions in the work of Heidegger, Merleau-Ponty, and Sartre, the best-known branches to sprout from Husserl's sturdy trunk.

Phenomenologists strive to understand subjective experience. They search for an underlying objective reality in subjective appearance. They care especially about consciousness. For our purposes, their message is simple. The first law of psychology—*all* psychology—is to *know what needs explaining*. You can't explain anything until you know just what you must explain.

What needs explaining then? The philosopher İlham Dilman tells us what is wrong with a famous mainstream philosopher who champions a computer-centered view of mind: "In his commitment to find a scientific explanation of consciousness he shows very little understanding of 'folk psychology,' treating its contents in a very cavalier fashion." ("Folk psychology" is commonsense, intuitive psychology.) Dilman continues: "What he needs is a *clarification* of the concept of consciousness, instead of an *explanation of it along scientific lines*."[7]

CLARIFICATION OF THE CONCEPT

The science *or* philosophy of psychology, writes the contemporary phenomenologist Eduard Marbach, requires "a systematic descriptive analysis of consciousness."[8] Where do we get this "descriptive analysis"? The phenomenologist Shaun Gallagher gives this answer: from "a methodologically controlled reflective introspection."[9] One must (methodically) *introspect*.

INTROSPECTION

Emergency! *911!* Most psychologists and philosophers hate the idea of introspection as if it were hell on wheels. Because most research-minded philosophers and psychologists are (after all) professional clear thinkers, their first impulse on confronting proponents of introspection is, naturally, to have them rounded up and shot. (Of course I am only kidding. They would have no legal means to have them shot. But the point is general: sentiment on the issue runs deep.) I will describe several other crucial information sources besides introspection. But let's not kid ourselves; you cannot study an inherently subjective topic like consciousness *without* introspection.

Phenomenologists remind us that when introspection is required, we must do it carefully, systematically—trying always to *see* or *intuit* (Husserl's words) the general laws revealed in our own experience. For every experience reflects the underlying laws of its domain—if there *are* any laws—as every rolling stone reflects the shape of the mountain slope. Through the peephole of my experience, I can see (if I am sharp enough) how consciousness must be structured in order to make my experience possible.

Introspection, however, is easy to misunderstand. Superficially, the introspectionist seems to say: "Here's how my mind works; I infer that some other minds (or many others, or *all* others) work the same way." But that is *not* the introspectionist proposition! (At least it should not be. It is not mine.) The introspectionist is a dance teacher, not a dancer. He demonstrates a move and then says, *you* try it! His goal is not to explain dancing by showing you how he does it. His goal is to explain dancing by showing *you* how to do it. Then you will reach your own conclusions. Thus the introspectionist goes in one leap from the weakest possible argument to the strongest—from a weak argument based on his experience alone to an irrefutable one based on yours. It is hard to deny the existence of phenomena you have seen in your own mind. "Until a person is able to fill up those concepts with their manifestations in his own life," writes Jonathan Lear, "his understanding of those concepts will be hollow."[10]

The "Little Room of Man"

The room with its view, the room in which we are confined, is an obvious metaphor for mind. King Lear, who comes to know the whole cosmos while staggering drenched over the moors through a screaming black thunderstorm, "strives in his little room of man" to come to grips with reality, with the thing itself. As usual, Shakespeare lays it on us exactly. The room with a view is a deep-rooted metaphor. "I live inside a skin inside a house. There is no act I know of that will liberate me into the world. There is no act I know of that will bring the world into me" (J. M. Coetzee, *In the Heart of the Country*). The novelist and memoirist V. S. Naipaul describes his proper topic: "The worlds I contained within myself, the worlds I lived in" (*Enigma of Arrival*). He lives within worlds within himself, worlds he himself has built: his little room of man.

The "little room of man," the room with a view, is equally a metaphor for the mind and for *phenomenal consciousness*—for subjective experience, that is, the moment-by-moment *feel* of reality. (For my purposes, "phenomenal consciousness" is the same as consciousness—"phenomenal" merely emphasizing that "consciousness" always means *subjective experience*.)

Consciousness is immediate, direct, and intimate—a sort of mental touching. You are conscious of something when your mind "touches" it and is touched *by* it. Consciousness is the *feeling* created by thoughts within your mind or by objects or events in the outside world.

We handle everything there is, in the outer and inner worlds, using the delicate cloth of phenomenal consciousness. Any interaction with the world is a thing of which we are conscious—therefore, a thing we experience, a thing we feel. The philosopher David Chalmers writes, beautifully:

Conscious experiences range from vivid color sensations to experience of the faintest background aromas; from hard-edged pains to the elusive experience of thoughts on the tip of one's tongue. . . . All these have a distinct experienced quality.

To put it another way, we can say that a mental state is conscious if it has a *qualitative feel*—an associated quality of experience.[11]

Charles Siewert is more direct: "'That noise sounded louder to me than the previous one'; 'I was visualizing the front door of my house'; 'it looks to me as if there is an X there.'"[12] Now the *quality* of consciousness (I've asserted) changes continuously and drastically as we move over the spectrum during the course of a day. But what *is* the "quality of consciousness"?

Here are some examples: "awake," "asleep and dreaming," and "out cold" are rough descriptions of three different qualities of consciousness. Subjective experience has a different quality when we are awake than when we sleep and dream. When we are unconscious, it's not there at all. "The first task of the science of mind," writes Hobson, "—to describe, define and measure polar states of consciousness such as waking and dreaming—has only recently assumed a serious status."[13]

Describing states of consciousness is the only place to start. Awake, asleep and dreaming, and unconscious are (however) only three states of consciousness on a continuum. We move from top to bottom of this spectrum or continuum nearly every day of our lives.

You are either asleep or not. But if we take one step back, to the moment just *before* sleep, we discover sleep-onset thought—a hard-to-know state of consciousness in which we tend to hallucinate as if we were dreaming. Back up one more step and we discover the free-flowing, associative thought that occurs when we are deeply drowsy. Take another step or two backward, up-spectrum—imagine moving from the bottom of a playground slide to the top—and we find the daydream- and fantasy-prone, easily distracted state of consciousness that is good for reminiscing but no good at all for systematic problem solving.

You have one personality, refracted into many states of consciousness by the prism of mental focus.

One more large step backward and we are midspectrum—no longer tired, but not fresh either; not at our sharpest, but ready and able to encounter the world. We can pay attention to a problem, work on it, focus. But prolonged focus is hard. Only when we have backed all the way up to the top of the slide, to our widest-awake, most energetic state of mind—not long after we've awakened from a night's sleep, usually—do we reach our best point for logical problem solving and systematic reasoning. A father teaches his children "not just to speak properly but to think logically, to classify, to analyze, to describe, to enumerate" (Philip Roth, *The Human Stain*). These are high-spectrum skills.

Looking at the process from its starting point—from the top— upper-spectrum changes are usually slow and subtle. We lose our edge, the energetic bite with which we strike into problems, our taut, focused attention, fairly soon. But then comes a long, slow relaxation, unfocusing until the mind starts to loosen up. Then we feel thought start to flow, and blow sideways in the breeze—not from one thought to logical next thought, but from one thought to illogical, *irrelevant* next thought, which somehow overlaps the previous one as thoughts do when we free-associate. Maybe the overlap is minor. Maybe we have no idea what it is. But when this happens, we have upshifted and are about to start moving faster. We will move from mental acting to mental *being*.

Memory changes too as we move down-spectrum. We live like chronicle composers bent over our parchments, looking up to scan the scene and back down to transcribe it into memory.

Memory is no literal chronicle: our memories can become confused, can be rearranged, can fade or simply be lost, or can be put down wrong in the first place. But we are like chroniclers insofar as *we live to remember.*

Experience is a body of memories we use (on purpose or implicitly) to guide us. But we don't experience an event merely by living through it. To experience an event, we must live through and *remember* it.

Surgeons will tell you that sometimes a patient is awakened briefly on the operating table when the procedure is almost finished, to make sure everything has been put back in place. But modern anesthetics and associated drugs ensure that no memories are laid down; the patient will never recall this little scene. Often he has a breathing tube stuck down his throat during the temporary reawakening—a sensation that is nearly unbearable in normal waking conditions. But it doesn't matter, because no memories are created, and nothing will ever be recalled; you'll never know unless someone tells you, and even then you might not believe it.

Did it ever happen?

Not in *your* experience. An experience is a memory. No memory means no experience.

Unremembered "paradoxical experience" (what we live through but don't remember) grows more important as we move down-spectrum. We recall little about the free association that accompanies drowsiness, little about sleep-onset thought, and dreams are famously hard to recall. Low-spectrum experience is often vivid and striking but "paradoxical," unreal, gone forever.

Experience at the bottom of the spectrum is hard or impossible to remember. Thus, it tends to erase itself: in the act of happening, it makes itself "unhappen." Some of the most vivid experiences we live through never *happened* to us, insofar as we cannot remember them.

Thus, as we move down-spectrum, we sink gradually (as the sun sinks in the west) into a deep reservoir of unrealized experience, *paradoxical* experience, which affects the mental world in subtle but profound ways. This is one of the stranger facts about mind. But it is an everyday fact, not a rare or exotic phenomenon. We must come to grips with it.

Usually, we oscillate partially up and down the spectrum several times over a day. Some people reach a comfortable spot for reasoning in the evening, but they still need to slide the whole rest of the way down-spectrum before reaching sleep. Different patterns hold for different people. The basic spectrum facts, however, apply to nearly everyone.

The "quality of consciousness" is, in short, a straightforward idea. Gradations in the quality of consciousness are easy to accept; we have all experienced them. A spectrum of conscious states is a simple idea. You would expect this spectrum to put in an appearance in chapter one of every introduction to the mind. Yet it doesn't. We are blinded by the subjectivity of mental reality, and the paradoxical experience of the lower spectrum.

Ultimately, too, we all suffer at times from the weakness that King Lear's nastiest daughter casually attributes to her old father: "He hath ever but slenderly known himself." That the mental spectrum should be largely unknown is just one more expression of the terrible trouble we have in knowing ourselves.

Long ago, an eminent professor of philosophy interrupted a lecture on Descartes to relate this story to the class: "A friend I hadn't seen for years told me, 'Do you know what your most obvious personal trait is? It's *this*.'" The trait itself remained a secret; we had to guess. The professor continued: "I couldn't believe it. It seemed absurd. Absolutely absurd. When I got home that day I told my wife, 'Can you believe what my friend described as my most obvious personal trait? *This!*' And my wife said, 'But of course.'"

Seeing things that are too close instead of too distant to make out clearly is one definition of philosophy and the philosophical method. "How hard I find it," writes Wittgenstein, "to see what is *right in front of my eyes!*"[14]

Authorities agree: we do not know ourselves. So it is no surprise, after all, that we do not know the spectrum that describes our own minds.

How Can We Know the Mind from Inside?

Ralph Waldo Emerson wrote an eloquent defense of what has become the grossly unfashionable practice of introspection. (Marilynne Robinson cited it in her indispensable *Absence of Mind*.) Emerson's advice to the young writer:

> In silence, in steadiness, in severe abstraction, let him hold by himself; add observation to observation, patient of neglect, patient of reproach, and bide his own time,—happy enough if he can satisfy himself alone that this day he has seen something truly. . . . For the instinct is sure, that prompts him to tell his brother what he thinks. He then learns that in going down into the secrets of his own mind he has descended into the secrets of all minds.

Husserl and the phenomenologists make it clear that to give up introspection is to disarm completely in the face of subjective experience. To understand, we must know our own subjective experience first. We must know it in a systematic, disciplined way. Our goal must be transcendental insight, in Husserl's sense: to see nothing less than *the shape of mind* in the small, local incidents we experience within our own minds.

But there are other, more important sources than introspection for this book.

I lean on some of the deepest thinkers and best-informed and most genuine experts mankind has ever known—the real authorities on the human mind. First, of course, there's Shakespeare. Second, and not even close—though far ahead of everyone else—comes Tolstoy.

Behind these, my choices are idiosyncratic to a point, but hardly surprising: Blake, Keats, De Quincey, Racine, Rimbaud, Hugo, Hölderlin, Büchner, Rilke, Kafka, Chateaubriand, Flaubert, Dostoyevsky, Proust, Jane Austen, Charlotte Brontë, Henry James, Ernest Hemingway, Vladimir Nabokov, Karen Blixen, Cynthia Ozick, J. M. Coetzee, V. S. Naipaul, and others. *These* are the people who know.

Wordsworth is in a class by himself, with his "mind beset / with images, and haunted by itself" (*Prelude*, Book 6). No one ever approached him in capturing the numinous light of early childhood. Freud is a special case too: a psychologist and philosopher of revolutionary depth and originality, with a great novelist's penetration into the stuff of life. He is one of the decisive, defining thinkers of modern times.

I tell my students that those who care about literature and mind must know the Hebrew Bible, Donne, Sterne and Jane Austen, Coleridge and Wordsworth, Proust and Kafka, Dostoyevsky and Tolstoy, and (of course) Shakespeare, to start. We are lucky to have a superb group of mind-mindful novelists at work today: Philip Roth and Martin Amis, Cynthia Ozick, Jenny Erpenbeck, John Banville, V. S. Naipaul, and J. M. Coetzee—to start.

Mind thinkers have often turned to literature. The phenomenologists make a practice of it; Freud and the psychoanalysts, even more so. But I must explain my special predilection for fiction. An eminent novelist or story writer must be a superb psychologist, must see straight to the bottom of human character; that's part of the job description. (We are talking *eminent* novelist.) Unlike a philosopher, psychologist, biologist, or technologist, a novelist ordinarily has no theory to defend, no psychological axe to grind. And we trust eminent novelists to take us where no one else can, into the *subjective reality* of a human mind. Novelists have no more data than anyone else. But they have psychological intuition that the worldwide community has evaluated and believes in.

Philosophers refer disparagingly to "folk psychology," the psychology or philosophy of the average peasant or serf. But what Shakespeare thought about the mind is not folk anything. It goes as deep as psychology can.

There is one additional source of information, a powerful one, about the subjective mind: the language we speak. Listening to language is crucial to a new subjectivist methodology. This is not the listening of analytic or language philosophy. *This* listening understands language as the hyperconcentrated, 180-proof, distilled essence of centuries' worth of thought and common sense about the mind.

Language is our handbook of common knowledge and common sense. Language is knowledge distilled, far beyond the concentration of any mere encyclopedia, into words, idioms, ways of speaking. "Feeling" is a synonym for "emotion." The mind can *wander*, can *daydream*, can *drift off.* You can *lose yourself, focus, snap out of it, see the world through rose-colored glasses*, or, on the other hand, *see red*. Information of huge value is captured here. Compared to the "smart" of human language, a smartphone is the smartest rock in the pile.

Now we are ready to pull on our boots, head into the mucky, sopping field, and (as carefully as we can) *observe*. Then think, and observe some more.

Two

Three Thirds of the Spectrum

Our movement across the spectrum is a simple operation with subtle, important details. We should have the freedom to observe the process from several angles. Suppose we are listening to an orchestra on a big bandstand in a park. We are circling around it, looking and listening from as many different viewpoints as we like. In this chapter I'll stop at three different points on the perimeter of the bandstand where the spectrum is, metaphorically, enshrined.

Everyone experiences the transformations that go with the spectrum. Everyone travels the whole distance top to bottom. But the road and the trip differ depending on you. "Top of the spectrum" means your best, clearest reasoning—which is radically different for a first-rate mathematician versus the average journeyman second baseman or brain surgeon. For some people, logical thinking consumes more territory than for others. For some, the middle zone of brisk, practical dealing is a comfortable place to linger. Our energy and focus levels determine our current spectrum position. But easy, natural work costs us less energy than hard labor. It costs less energy to linger where things come easily. So it is natural to draw such moments out. Your own spectrum, and your own method of moving across it, enter into the delicate balance between a universal pattern on the one hand and personal qualities and habits on the other.

View 1: The Transformation in How We Make Sense of the World

In this first view, as we move down-spectrum, mental activity changes—from largely under control to *out* of control, from thinking on purpose to thought wandering off on its own. Through it all, the mind has a powerful urge to make sense of the universe—to understand, that is, how the countless pieces at countless scales relate; to assemble enough of the jigsaw puzzle to see the slowly emerging big picture. Up-spectrum, the mind pursues meaning by using logic. Moving down-spectrum, it tends to pursue meaning by inventing stories—as we try to do when we dream. A logical argument and a story are two ways of putting fragments in proper relationship and guessing where the whole sequence leads and how it gets there. This is the logic-versus-narrative axis.

> *Spectrum Law:* The mind is in business to make sense. Up-spectrum, it makes sense by making logic. Down-spectrum, it makes sense by creating stories.

Up-spectrum, with mental focus high—we might imagine the thinker atop a mountain on a bright early morning, with a clear head and commanding view—we are willing and able to pay close attention to what's happening around us, to make plans and solve problems. We rarely think logically in any precise or formal sense, but we lay down mental tracks step by step to our goals.

As energy dips and freshness fades, and our thinker makes his way down the mountain trail, we attend to business, keeping our minds on our work and surroundings. But we concentrate in shorter bursts and attack problems using less logic and more experience (what *usually* works?), and we grow increasingly distractible.

We say *our minds wander*, go off by themselves; memories, day-dreams, and fantasies lead us away from the task at hand. Further down-spectrum—it's evening, maybe getting late—daydreams can engross us completely. (During a long walk, Wordsworth stops to rest in a forest grove, lies on the ground, and daydreams. He is "lost / Entirely, seeing nought, nought hearing" [*Prelude*, Book 1].) And eventually we *drift off to sleep*. "Drifting" means motion we can't control.

Reason and logic are ways of assembling thoughts to bridge the gap between where we are and where we want to be. Reasoning requires mental focus and concentration, and our ability to bring it off decays as we move down-spectrum. But another strategy for building thought sequences emerges. A daydream or a fantasy is a narrative—often short and weak on plot, but a narrative. In dreaming, we spin longer stories. It can be hard to turn a collection of memories and ideas into a story, just as it can be hard to build a logical path from premise to goal. But minds are in business to make sense. Logic and narrative are different ways to go about the same task—opposite ends of one spectrum.

The scientist explains the origins of the universe with a logical argument. The religious believer tells a story. (When I say "believer," I mean Jew or Christian—the only religions I know sufficiently to speak of.) Only the logical argument has predictive power. Only the story has normative moral content. Only a fool would pronounce one superior.

STORYTELLING

Notice how satisfying stories are at the end of the day. The bedtime story is a practical device for calming children, and it works well in part because we are in the mood for hearing *and* telling stories when we are down-spectrum. Jane Austen's Emma was required for weeks on end to tell exactly the same bedtime story to her visiting nephews, who were "still tenaciously setting her right if she varied in the slightest particular from the original recital" (*Emma*). They don't want to *know* the story; they want to *hear* it, as one hears a song. Story hearing is the perfect lead-in to dreaming.

Some adults read themselves to sleep. Proust did, and he would routinely dream himself into the center of the story he had been reading. Story and dream would blend perfectly. "I hadn't stopped reflecting, when I fell asleep, about what I had just been reading, but these reflections had taken a slightly peculiar turn" (*À la recherche du temps perdu* [*In Search of Lost Time*]). Stories are plainly easier to remember than other forms of information. "There are astonishing feats of memory. . . . Illiterate bards sing long epic poems in remote Yugoslav taverns; . . . oral historians in Liberia remember the histories of whole clans and tribes. Their performances are impressive. How can they remember so much?"[1] One reason is that stories are made for remembering. By following the natural contour lines and force fields of character and plot, they yield narratives that are not merely engaging, that we like to hear, but that suit our memories the way native flora suits the landscape.

The mythical Scheherazade, the beautiful maiden who sleeps with the king of Persia despite his bad habit of beheading each new bride at daybreak, is the embodiment of a basic urge. Each night she tells the king a fascinating story and breaks off at dawn, partway through—and he spares her, holds her over just long enough to finish her story the next evening. Whereupon she immediately starts another story, is held over again, and continues through the famous cycle of the thousand and one nights. It is almost as if she is dictating the king's dreams straight into his inner self, his passive, wide-open, parched-for-stories memory. She is the goddess of the lower spectrum.

THEME CIRCLES

I am using the terms "story" and "narrative" in a slightly nonstandard way. The raw stories our minds invent have no conventional narrative unity. They have *thematic* unity. A dream is a loose bundle of scenes that change abruptly and obviously don't sustain a plot or characters as any ordinary folktale or bedtime story might. A dream is more a set of anecdotes (sometimes bizarre) fastened together, with no more narrative consistency than a charm bracelet. But when we look carefully at those anecdotes, we usually discover that they are all *about* the same thing—all have the same theme. I'll call this a "theme-circling" narrative.

The theme is never stated explicitly; the many segments of narrative are like many segments of one circle without its center point marked. But the whole circle *points to* the center, making the center clear even though it is not marked. And the theme of these narratives, these circles, is usually quite clear when we stop to think. We might think of such a narrative as disorganized, without start or finish—like a mature novel by the eminent V. S. Naipaul. Naipaul himself writes (in a novel): "Life doesn't have a neat beginning and a tidy end. Life is always going on. You should begin [your novel] in the middle and end in the middle, and it should all be there" (*Half a Life*). Of course, such a story *is* organized, by the character whose life is under discussion. His personality is the theme of this theme circle.

A theme-circling narrative, a string of anecdotes (in effect) that reads just as well backward as forward, is in many ways less sophisticated than a carefully plotted work that builds to a climax. Yet the theme-circling work has its own sophistication. It points implicitly but with perfect clarity to a theme that *isn't there*—that is never stated explicitly—for a good reason. The theme is probably an image or emotion or both: an image charged with emotion, like a sweater with static electricity. Such themes are not hard to conjure up indirectly, but they are hard to describe explicitly.

An ordinary narrative has themes too, naturally. Its themes might be images or emotions or anything. But the repetitions in a theme-circling narrative, each segment pointing toward the same center point, guide readers or listeners to the theme. Awareness of the theme grows naturally in the reader's or listener's mind as the theme circle progresses. The circle has its own technique, its own nuances, and must be taken on its own terms.

Narratives are created in several ways in the lower spectrum. We tend to store rough notes, not multimedia scenes, in our memories. To retrieve a recollection, we *reconstruct* the full scene from our notes—a reconstruction based on similar memories, but also on what seems right, what fits. "Re-creation is the key term here," writes the psychologist David Foulkes. "Conscious episodic recollections are momentary constructions out of information left behind by past experience"—in other words, out of notes or partial recollections of past experience.[2]

This inventing of appropriate detail is a quintessential *storytelling* task. The first manifestation of our switch from logical to narrative thought is this conversion of rough notes into full recollections. We are like reporters transforming notes into news stories.

FROM LANGUAGE TO IMAGES

Dreaming is *storytelling in pictures*. "All of us try to make sense of our lives by telling our stories," writes the psychoanalyst Stephen Grosz.[3] (Of course not every spectrum point is equally hospitable to the telling of stories.) "Visualization of the fluid sort that we find in our dreams," adds Foulkes, "is a *thought process*."[4] He is echoing Freud: "At bottom, dreams are nothing more than a particular form of thinking."[5]

Why do we think in images and tell stories at the bottom of the spectrum?

Language and pictures are not simply two ways to express ourselves. For one thing, we teach children how to speak and write but not how to draw; in one sense, the deck is stacked. Nonetheless, language is in many ways natural up-spectrum, where we want precision and conciseness and often rely on abstraction. Pictures do well down-spectrum, conveying concrete richness of detail and (sometimes) the nuances that are essential to emotion and atmosphere.

Nabokov wonders whether a memory of his first girlfriend will "survive captivity in the zoo of words" (*Speak, Memory*). Although he can write sentences like *that*, he is still uncertain whether words will convey the sensory richness and mystery of memories so heavy with emotion that they arc backward like long shoots of grass touching their tops to wet ground after heavy rain. Nabokov has a strong bias and talent for thinking in pictures—but he is, of course, a master of words too. (This combination is more the rule than the exception.) Less widely understood is the power of pictures to capture elusive abstractions that one intuits before clearly understanding them. And of course, vision is our primary sense. When we remember the past, we usually remember in pictures. And we will see that dreaming is, first and foremost, *remembering*.

Images have another good property as a basis for thought: it's easy and natural to combine them to make complex but vivid composites. I can take a memory of walking down the sidewalk, add a snowstorm, subtract people and buildings, add the sound of a fast-approaching police car—a wide range of options. "A whole Essay might be written on the dangers of *thinking* without Images," writes Coleridge, the poet-psychologist-philosopher (*Biographia Literaria*).[6] Nonetheless, in fairness, as I discuss later, the real alternative to language as a form of communication is the body itself. Language is natural to logical, up-spectrum communication; wordless gestures, facial expressions, tone of voice, ways of standing are natural to emotional, low-spectrum communication.

We see children demonstrate in their own lives something like the hypothetical emergence of logic from stories. Children love stories but gradually make room for logic and nonfiction too. We see a maturing child able to reach higher and higher points on the spectrum month by month, increasingly able to manage logic and abstraction. We see signs of the same transformation in the coming of age of civilizations. Human literature began by telling stories and hasn't stopped yet. But along the way we have developed a propensity for history too.

We have developed an interest in objective truth. Our growing science-centeredness is one coarse index of the ever-more-focused thought styles dictated by fashion and preference in a maturing culture. The widespread beliefs in the modern West, fuzzily defined or just implicit among the believers, that legal worldviews are more sophisticated than religious ones, and sexual worldviews more than romantic ones, are more symptoms of up-spectrum life. In each case, the spectrum is just one tree in a large grove of explanations. But it's a sturdy tree, and that's another story.

Let's move to a different viewing angle and watch from there. Here is a second axis of change—subtler, but equally important.

View 2: The Transition from Acting to Being as the Main Focus of Mind

There are two main facts about the mind: acting and being—what the mind *does* and how it *is*. (Aristotle made an influential distinction, in discussing the mind, between action and passion, but my dichotomy is different.) The first pole, mental *activity*, is, broadly speaking, the deliberate manipulation of conscious mental states: thinking about something, or *reflection*. The second pole, mental *being*, is sensation and feeling: *feelings*, physical or mental. As we slide down-spectrum, our mental center of gravity moves from acting to being. From thinking about to sensing and feeling.

Spectrum Law: Up-spectrum, the mind is dominated by doing. Down-spectrum, it is dominated by being.

We never *don't* feel. To be conscious is to experience, to feel. But as we move down-spectrum, feeling and sensing move from the periphery of mind to dead center. Let me explain.

ABOUTNESS DISSOLVES

Much of mental life has a topic: my car, your car, where to park, Mike and Erica, tension in the European Union, peanuts. And much has no topic: I am dizzy with happiness, in pain, afraid to move, enraged and bitter, smugly triumphant. "Smugly triumphant" is not the topic of my mental life, although it could be. I could be pondering smugly triumphant, but I'm not; I *am* smugly triumphant. I might be giving a lecture about fear, or (on the other hand) I might *be* afraid. Or I could be both: aware of a feeling called "fear" while I experience fear. But I can perfectly well be afraid yet unaware of the fact.

It's not that I'm deluded, not that I think I'm cheerful and carefree. It's just that I am wholly, deeply engaged in the sensation or experience of fear and have no attention left over for anything else—not for awareness of my sensations, or self-awareness generally; not for whistling a happy tune; not for anything. By "sensation," I mean the many subtle body feelings associated with fear; by "experience," I mean those feelings *and* my immediate emotional responses: perhaps anxiety, jumpiness, mind racing—or maybe cool resolution.

Whether I am merely aware of my sensations or feelings, or wholly engaged in *having* them, is partly a matter of how strong they happen to be. But it also depends on where I am on the spectrum. The act of thinking about, of stepping back and examining myself and my sensations, comes more naturally up-spectrum than down. Down-spectrum I tend, increasingly, not to think about, but just to *be*.

Aboutness dissolves—mental topics in general dissolve—as I move down-spectrum from action to pure being. In philosophical language, "intentionality" dissolves. Intentionality is the quality of aboutness, of *referring* to something. "I believe it is about to snow" is an intentional state; the *belief* is *about* something—namely, the weather. Intentionality grows dilute and finally dissolves completely as we move down-spectrum.

Believing it is about to snow is an intentional state; being depressed is *not* an intentional state. Being depressed is about nothing. It is just *a way to be*. You are depressed in roughly the sense that a petunia is purple. Purple is how that petunia is. Depressed is how you are. (That depression is not intentional makes it no less important.)

The pleasant coolness of my forearm is not *about* the cool, fresh breeze. My nostalgia or my anger (say) have causes, as the coolness of my forearm does, but they are not intrinsically *about* anything. Anger and nostalgia are just ways a person can feel. They are *ways to be*.

Intentionality was introduced to modern philosophy in the mid-nineteenth century by the German philosopher Franz Brentano, who got it from the medieval scholastics, who found it in Aristotle. Brentano, however, used the old idea in a new way: to distinguish states of mind from everything else in the universe. All states of mind and nothing *but* states of mind are intentional, said Brentano. No mere physical state, on the other hand—no state of a tree, planet, photon, tomato—can be intrinsically *about* anything. It was a brilliant and useful observation, half true. *Only* states of mind can be intentional, but it is widely agreed that some states of mind are not.

Brentano's answer was wrong, but his question—what is specifically mental about the mental?—was a milestone. What makes a state of mind different from all other states? The answer, I believe, lies in subjectivism. All and only states of mind are two-faced, two-sided, double states. (The eminent philosopher Thomas Nagel has said so, in a different way.) Your mental states have an outer, objective side that anyone can see, and an inner, subjective face that is visible to you alone. Everything else in the universe, so far, is one-sided, with objective properties only. Everything else has only outer; minds alone have outer *and* inner.

Rainer Maria Rilke puts it vividly in the eighth "Duino Elegy": "With every eye, Creation beholds the wide-open world. Only *our* eyes are turned backward. . . . We can know what is out there only from the animal's face." Animals only look *out*, never in, and they have no human-style selves to worry about. We, on the other hand, can never forget or shrug off the room we are locked in forever.

Brentano was wrong about mental states but right that thought is always *about* something, and thinking about dominates the upper spectrum. But as we move down, aboutness dissolves and mental life comes gradually to center on our own pure being. Watching the spectrum run backward, from bottom to top, we can almost imagine that we are seeing evolution gradually carve out a place in the mind for reflection, for *thinking about*, as our capacity for abstraction slowly emerges.

In terms of aboutness, feelings and music are analogous. Music can be *inspired* by something—the departure or return of one's beloved, Shakespeare's *Tempest*, a gentle brook in the countryside. But it can't be *about* any of those things. Music is not capable of aboutness. The notes of the scale can communicate emotion but not information. And emotions can't be about anything either.

THE SPECTRUM FROM THE STANDPOINT OF EMOTION

Up-spectrum, we keep our emotions on a short leash, because emotion disrupts thought as catnip disrupts cats. Emotion is thought-nip. Up-spectrum, we rarely give way to emotion. We focus on our plans, goals, surroundings. Early mornings are rarely the time for storms of rage or despair. Nor are we normally at our wittiest or most engaging at breakfast. ("'Aw, hell!' I said. 'It's too early in the morning'" [Hemingway, *The Sun Also Rises*].) Down-spectrum, emotions gradually emerge. Daydreams and fantasies are usually emotional. Many dreams are highly emotional. And the most intensely emotional experiences our minds ever concoct happen in dreams. But there is more.

Approaching bottom, in the fullest bloom of pure being, even our sense of self is shouldered aside. We *forget ourselves*, lose our selves. We become mere *experiencers*. We encounter selfless states of pure being.

States of pure being are dangerous and can terrify us: we are wide open and have no defense against nightmare. Such states can also bring us to the very lip of volcanic euphoria—or flood us with deep contentment, or make us (briefly) one with the universe and show us a spark, gleam, glimpse, or lightning flash of what is meant by "God."

AT THE BOTTOM OF THE "EMOTIONS SPECTRUM"

At the spectrum's bottom we are asleep, or almost, or awake in a daze that resembles sleep—although we might be active and encountering the world. Near the bottom, strange states happen. Let's consider the "selfless state of pure being."

States of pure being might seem, if nothing else, *simple*. Feeling or sensation dominates the mind. Period. But these states are not simple. They happen and *unhappen* simultaneously. They challenge our sense of the real. For this if no other reason, philosophy of mind cannot ignore them, as it likes to.

These states are paradoxical insofar as we start to lose our sense of self *and* the capacity to form sound memories at about the same time. We don't know exactly how a lost or weakened sense of self hurts our memory-making ability—if it does. And it seems to. But unless we focus attention on a new memory, *think about it*, it stands little chance of surviving. Self-awareness being shouldered aside suggests that awareness in general has been pushed out of the way. We are caught up in pure being, sitting back and smelling the roses, not pondering or reflecting, *not* making solid memories. We have trouble finding, as well as forming, these low-spectrum memories. We encounter the problem of overconsciousness.

"Overconsciousness." Psychologists have never even heard the word. (Of course not! I just made it up.) But this phenomenon without a name has always been important. Overconsciousness happens when an exclusive focus on experience shoulders aside our usual sense of self, self-awareness, reflectiveness—and makes memory work badly or fail completely. It happens routinely down-spectrum, but it can crop up anywhere when abnormal events surprise us.

In overconsciousness, sensation or emotion fills our minds to the brim. The principle is simple. We have just so much attention to give. If we use it all on one thing—one sensation, one experience— we have none left for anything else. We can watch one place or person carefully, two at the same time less carefully, ten even less carefully. Unfortunately, we cannot increase our supply of attention to match increasing demand.

In overconsciousness, one powerful, coherent group of stimuli makes it hard to do anything else with our minds—to reflect on what's happening, to track and remember other simultaneous events. Imagine looking straight at a lightbulb that flashes so bright you can't quite see it. You could also live straight through an event that's so "bright" you don't quite experience it—because you don't reflect on, don't *remember* it in the normal way. Flash burn is something like overconsciousness ("consciousness burn"), and we might not quite be able to understand dreaming—or earliest childhood—without understanding overconsciousness.

TESTIMONY: OVERCONSCIOUSNESS, "CONSCIOUSNESS BURN"

How does overconsciousness occur? How is it described, or *not* described?

The eyewitness to an accident in J. M. Coetzee's *Age of Iron* says this: "Time seemed to stop and then resume, leaving a gap: in one instant the boy put out a hand to save himself, in the next he was part of a tangle in the gutter." Perhaps the experience is uncommon, but we recognize it. Coetzee's account rings true. Ernest Hemingway describes a serious artillery wound: "A roar that started white and went red and on and on in a rushing wind. . . . I felt myself rush bodily out of myself and out and out and out" (*A Farewell to Arms*). This is a novel, but Hemingway himself experienced just this type of war wound. Here we read experience reconstructed after the fact. But the "on and on," the "out and out and out," sound like mental blanks, *memory* blanks.

I can experience something, as I have mentioned, without being *aware* that I am experiencing it. My mind is full of the fresh, raw experience itself—of being thrown through the air (say) by an explosion: I *sense* many things—the rush of air, the feel of falling or flying, disorientation, the ground in a blur—and have no mind left over to reflect. But without self-awareness or reflection, with no interpretation of what is happening, I have no basis for memory—except for memory of the sensations themselves. I can't remember what I never knew.

Thomas De Quincey says: "Rightly it is said of utter, utter misery, that it cannot be *remembered*" (Salaman, *A Collection of Moments*). In saying so, De Quincey cites Coleridge: "I stood in unimaginable trance / And agony, which cannot be remembered" (*Remorse*). Philip Roth in *The Human Stain*: "Delphine is so stunned that, later, she does not remember putting down the receiver or rushing in tears to her bed or lying there howling his name." In *The Zone of Interest*, Martin Amis writes with terrible acuity that newcomers to Auschwitz saw sights they simply could not absorb, that went unremembered because they were too horrible for the mind to take in.

In Esther Salaman's words: "One seldom remembers a moment when one's whole self goes into a passionate or violent action." No mind, no *self*, is left over for self-awareness or creating memories. This is "consciousness burn."

Overconsciousness is easy to recognize in traumatic circumstances. But it happens in other ways too. It happens nearly every day to nearly everyone, at the bottom of the spectrum.

Some personalities fall naturally into daytime overconsciousness. "Nikki was explosive, crazily emotional, and could do and say bizarre things and not even remember them afterwards" (Philip Roth, *Sabbath's Theater*). Crazy, explosive emotion monopolizes your attention. Georg Büchner's manic-depressive hero Lenz recounts an experience he had during a hike through the mountains:

> The storm would drive the clouds downward, and rip into
> them a light-blue sea, and the wind would hold off and then

murmur upward out of the deep gorges, from the tops of the fir-forest, like a lullaby, like a peal of bells . . . it *made a tearing in his chest—he would stand gasping, body bent forward, eyes and mouth wide open.* (Georg Büchner, *Lenz*; italics mine)

This is a powerful reaction to the beauty of nature. But we also notice overconsciousness in portentous yet perfectly normal childhood experience. "Remember what April was like when we were young," writes John Banville, "that sense of liquid rushing and the wind taking blue scoops out of the air and the birds beside themselves in the budding trees?" (*Ancient Light*). The child's normal experience sounds almost like the adult's highly *abnormal* encounter with artillery fire. A child's mind is biased toward the spectrum's low end, toward vivid sensation.

As a child, J. M. Coetzee believed that the young colored children of his South African town—lighter-skinned than "natives," in the old-time South African view—would be right to hate him for his birthday treat in a pastry shop that they couldn't afford. They gathered at the window and watched. *Yet their faces showed no hatred.* Puzzle. Coetzee doesn't explain, but he supplies all the data we need. Watching through the window, the small children "are like children at a circus, drinking in the sight, utterly absorbed, missing nothing" (*Boyhood*). When sensation and emotion flood the mind, there is no room for anything else, not even natural anger or hatred.

Karen Blixen writes in *Out of Africa* of her East African, as opposed to European, friends: If you travel into town together, go into a building to do an errand, and leave the African behind for a while, "he does not try to pass the time then, but sits down and lives." Being as opposed to doing. "Boredom is a sentiment not available to the Hottentot," says Coetzee's narrator in *Dusklands*, with contempt for supposedly stunted Hottentot intelligence. But to take satisfaction in just living, in pure *being*, to have no concept of boredom—these are lower-spectrum attitudes whose resplendence we can barely guess.

"Ready for what I don't know," says a Philip Roth character. "Not just making love. Ready to *be*" (*The Human Stain*).

PURE BEING AND THE DREAM

Overconsciousness happens to us nearly every day. We all reach this exotic state routinely. We merely slide and tumble like children down the slope of the spectrum and find overconsciousness at the bottom, as consciousness itself burns brighter and brighter in the pure oxygen of the spectrum's lower reaches.

Proust speaks of "all these mysteries that we believe ourselves not to know—and into which we are in reality initiated almost every night" (*À la recherche du temps perdu* [*In Search of Lost Time*]). Every night we experience, in dreams, sensation or emotion so vivid as to occupy our minds almost completely and leave us no space, or not much, for self-awareness or reflection or making memories.

"They slept again . . . , dreamlessly, or so they believed" (Cynthia Ozick, *Foreign Bodies*). We often believe such things.

Hallucinations come in many varieties. Some are psychopathic symptoms. But imagine an "ordinary" dream-type hallucination in which a normal scene appears before you, perhaps with motion and sound—just like life. You would expect such an experience to be enveloping, to command your attention. Compare the recollection of a huge red-orange maple in fall to a hallucination of the same maple. When we hallucinate, we don't just recall the memory; we *reexperience* it. We reenter the experience instead of merely inspecting it from outside. A hallucinated recollection is clearly more involving, enveloping, and attention-grabbing than a typical recollection.

Ordinarily, the hallucination overwhelms the hallucinator—who does *not* reflect, "I am now hallucinating. This is not real." Such reactions can and do happen, but they usually don't. Ordinarily, we are taken in.

Hallucinating usually means dreaming. While we dream, we do *not* tend to reflect about the dream (or hallucination) that is happening right at the moment. Sheer experience—pure emotion and sensation—overwhelms us, and we approach a state of pure being.

Thus the sleep researchers Inge Strauch and Barbara Meier write that, in dreaming, "we deal predominantly with events of the moment" (the ones in the dream), and "these demand our total attention."[7] *Total.* It is rare to ponder car insurance while you are dreaming about a magical faraway landscape. It is rare to ponder *the dream itself,* to *reflect* on what you are dreaming, to consider the fact that "I am now hallucinating." The dream demands total attention. There is no room (or not much) for anything else, and thus we reach pure being, a pure way to be—overconsciousness, or almost.

"As is typical of most dreams," writes Hobson, "I am so involved in the scenario that it never occurs to me that I am dreaming."[8] Consider similar statements: I am so involved in what's happening, it never occurs to me that I am watching a movie, or playing the piano before a large impromptu audience, or charging the enemy while I fire my weapon and bullets suck the air all around me. A hallucination or dream is, by its very nature, a commanding, involving experience that leaves little mind left over.

Overconsciousness rules out self-awareness. We become pure sensation, pure emotion. In this realm of pure being, the self dissolves.

That fact is reflected in the well-established nature of the self in our dreams. It's been understood for generations that our dream selves are nearly always at the center of the dream action—but are radically different from our normal selves. They are hollowed out. They are unreflective crypto-zombies. We accept all sorts of absurdities in our dreams. We don't say, "*That* can't happen! What's going on here?"

We need a "self" character to organize the dream story, but the dream self is so strange that it almost tells the story of overconsciousness all by itself. Hobson, again, writes that our dream selves "can't keep track of time, place or person, and can't think critically or actively."[9] David Foulkes wonders, "How can our awareness of *dream events* be so vivid, while we are so little aware of what our *minds* in fact are up to?"[10] Because all our attention goes to the spectacle of the dream, none is left over to observe the dreaming mind. The first of Hobson's "cardinal cognitive features of dreaming" is "loss of awareness of self (self-reflective awareness)."[11]

THE EMOTIONS AXIS, IN SUM

As we move down-spectrum, we gradually lose control over thought; we grow passive. Mental states grow more vivid and enveloping. We daydream, fantasize, get lost in a maze of free association, and then wander dazed into sleep-onset hallucinations. At each stage, less attention goes to thought—to mental doing, organized mental *acting*—and more to experience, to pure being. Dreaming is the natural end point.

Overconsciousness is also a contributor to the other great mystery of forgetting in normal psychology—"infantile amnesia," Freud called it. "Days / Disowned by memory," writes Wordsworth (*Prelude*, Book 1). Before some age, usually between two and four years, we recall nothing. Might infancy and early childhood be states in which sensations overwhelm us, leaving us unreflective, not self-aware?

Mind science has not yet captured the special phenomenal status of the lower spectrum, the strange kind of experience we meet there.

The philosopher Brian O'Shaughnessy, however, describes something much like *some* of this axis in a Freudian key. He writes that the thinking process is "the central phenomenal process of mind," and that thinking covers a sequence or *spectrum* of processes, including "creative thinking, the day-dream and the dream." "What has impressed me," he says,

> is that a continuity links the *most* self-determined of these [thought processes], attending, and the *least* self-determined, dreaming; and that as one moves along this spectrum, *pari passu* the will fades from the scene.[12]

In other words, our ability to exercise deliberate control over these thought processes fades. I don't know of other references in the literature of mind to a *spectrum* of mental states, but whether or not they use the word, the Freudians are most likely to think the thought.

Our experiences grow more intense, our memory spottier, as we move down-spectrum. It's hard to estimate how much of our lives we spend dreaming. But that part *plus* the unremembered periods that lead into dreaming equals the segment of our own experience that is not part of our lives. Toward the bottom of the spectrum we take the most precious of all substances, conscious experience, and pour it out on the floor. Only by understanding what happens during those periods, and the indirect ways in which those occurrences affect us, can we reclaim any of this precious spilt life.

View 3: The Transition from Outer to Inner Field of Consciousness

As we move down-spectrum, we become increasingly inner-focused until, drowsy and with mind wandering, we all but ignore the outside world. Then we find ourselves *beneath* the waterline, asleep and dreaming, sunk into our own memories, in a strange private world where external reality is just barely audible when it shouts.

As we move down-spectrum, we sink into ourselves, into our own minds. "I deepened myself," says Coetzee's narrator in *Dusklands*, "into a boyhood memory of a hawk ascending the sky in a funnel of hot air. The stillness of the wagon awoke me." Notice that, in "deepening himself," of course he falls asleep.

As this shift in focus happens, the very nature of remembering changes. Down-spectrum, we can remember things, especially details about the past, that were invisible or inaccessible at higher spectrum points. (It's no accident that Coetzee's narrator should plunge into a *boyhood* memory.) This third axis centers on consciousness of the outer versus inner world. This is the outer-to-inner consciousness axis.

> *Spectrum Law:* Up-spectrum, we live in the present. Down-spectrum, we tend increasingly not merely to *recall* but to revisit, briefly to *reoccupy*, the past. We sink into ourselves, out of the present. Yet, paradoxically, the only moment we ever can experience is *now. Experience*, phenomenal consciousness, *ties us irrevocably to now.*

Of course I can think consciously about the past or future. But I can *experience* only the present moment and no other. Even when I reexperience the past, I reexperience it at this moment, *now*.

What seems, perhaps, like a change from a now-centered to a then-centered view as I move down-spectrum is, in fact, a change from a space-based to a time-based view. My awareness of time, now *and* then, increases as I move down-spectrum. ("Everyone tends to remember the past with greater fervor as the present gains greater importance" [Italo Svevo, *Zeno's Conscience*].) Recall Kant's view of space as the intuitive organizer or *form* of the outer world, time of the inner.

Tumbling back into the past is a side effect of a crucial daily change in the mind's workings. Thoughts and memories can be conscious or unconscious. It's generally accepted that the conscious and unconscious minds exist side by side, although there are deep disputes about the unconscious mind's role. We speak about mental events *under control* and *out of control*: we do some mental acts deliberately; others happen to us. We can decide to daydream. We cannot decide to dream—at least, not ordinarily. Clearly *under control* versus *out of control* has to do with the conscious mind versus unconscious mind. A conscious mental act is under "my" control—"my" meaning, as always, *my conscious mind's*. An unconscious mental act happens *to* me, out of my control.

Up-spectrum, the conscious mind is in charge. The unconscious gradually asserts itself as we move lower. During the same interval, memory's behavior changes. Naturally, all these changes are connected.

Up-spectrum, we use memory mainly as an information source. We want data, not recollections. What's the name of the person who just said "Hi"? Will there be parking in the next block? Why is Jill sulking? Often the information is ready and waiting. Sometimes it must be created on the spot. Relevant recollections are fetched; information is extracted, like metal from ore. The *information* matters, not the recollections.

As we move lower, memory takes on its more characteristic function. I am reminded of scenes and events that emerge from memory and allow me to think back, reminisce, put things in perspective.

And the *quality of time* changes. Our focus moves from outer to inner. Time is no longer the ticking clock; it is something you create, you exhale—as if you were a pianist at the keyboard, improvising time. It is something *you* make, to accompany your life. Your attention is off the outer world and its clocks, and *on* the natural, varying beat of your inner world—your sensations and feelings, and the moods and memories that follow.

In the upper spectrum, you dance to the music. At the bottom, you are still dancing, but the music is inside your head, setting the rhythm and pushing time forward all by itself. In the upper spectrum, time is clear, tasteless water; in the lower, it has character— *just* the character you give it. Hence Nabokov's wonderful phrase "the texture of time," repeated often in his masterpiece *Ada*. In the right circumstances, time is a fabric whose texture varies. For the poet Paul Valéry, looking out to sea, *le temps scintille*, "time glitters" ("Le cimitière marin" ["The Cemetery by the Sea"]); time, like the sea, is a mere background that colors everything else.

BACKWARD IN TIME

Turning away from outer to inner dims the lights of the outside world—closes the curtains—and the inner world, inside our own room, grows correspondingly brighter. The growing vividness of memories, eventually turning hallucinatory, is partly the consequence of darkness everywhere else. The increased importance of emotion as we move down-spectrum is also, in part, a result of lowering the lights on the external bustle that fills our minds up-spectrum. To feel emotions properly—the nuanced valleys and not just the peaks—we must be attuned to the quiet inner world. Jane Austen tells us so. In the famous conversation leading to the plot climax of *Persuasion*, Anne tells Harville, speaking of early-nineteenth-century women generally: "We live at home, quiet, confined, and our feelings prey upon us." She is explaining why women are more faithful than men. For one thing, she says, we have less to distract us from our own emotions. We are less able to escape our own feelings.

When we shift emphasis from the outer to the inner field of consciousness, we sink into ourselves. Sinking into ourselves, we sink into the past.

Our subjective selves are objects built in time—as a road or a tree is an object in space. A human life is a kind of village in time, with a thousand small buildings dotting the temporal landscape, most clustered in early childhood. Each holds one moment that is hidden away in our minds like the grottoes, secret lakes, and diamond-clear underground pools in silent caverns of an enchanted island, of Xanadu. Most we will never see. No one will. Yet some *will* be revisited.

Our memories (especially but not only of early childhood) are sometimes not merely recalled, but reinhabited. Our uncanny capacity to reenter memories of past time—not merely to know them, but to *reexperience* them—is a defining aspect of mental reality. We can remember a trip to the beach by inspecting the memory from outside, as we normally do—or we can revisit, reenter, reexperience it, as we do in our dreams. "How small the cosmos (a kangaroo's pouch would hold it), how paltry and puny in comparison to human consciousness, to a single individual recollection" (Nabokov, *Speak, Memory*).

Those eternally preserved moments of past time are locked doors inside us, and if we dive deep enough, we can find the doors, pick the locks, and reenter the naked past just as we first lived it. In fact, most of us do so nearly every day—although afterward, we forget all about it. Pure being makes for low-quality memories, or none at all.

STRANGE STATES ON THE WAY TO SLEEP

As we approach sleep, our minds pass through strange states that we almost never recall.

Past the hallucination threshold (in fact, we might pass it, retreat, and pass it again several times), we are hallucinating, as when we dream. But the distortions of actual dreaming have yet to begin. So we reenter old memories as they happened, not in their mixed-up dream forms. In effect, we have been led, in secret, backward through time into moments we thought were gone forever. We can't *participate* in these moments (which, after all, have already happened), but we can observe firsthand as long-lost people and places reappear.

Sleep-onset thought is rarely mentioned in any sort of literature, but Charlotte Brontë knew all about it. "Sometimes I half fall asleep when I am sitting alone, and fancy things that never happened," her narrator says in *Jane Eyre*. To "half fall asleep" is a good way to describe sleep-onset thought. The narrator continues:

> It has seemed to me more than once when I have been in a doze, that my dear husband, who died fifteen years since, has come in and sat down beside me; and that I have even heard him call me by my name, Alice, as he used to do.

Notice that this character speaks first of imagining "things that never happened." What she is describing, however, is not a fantasy but a *memory*, recalled, *realized*, with hallucinatory vividness "in a doze"—on her way to sleep.

Coetzee has the same precise, polished knowledge of sleep onset that he has of every other square inch of the human psyche. A sudden flood of memories at the theater in the evening: "As if he has fallen into a waking dream, a stream of images pours down, images of women he has known on two continents. . . . His heart floods with thankfulness. Where do moments like this come from? Hypnagogic, no doubt" (*Disgrace*)—that is, "leading to sleep," another term for sleep-onset thought.

THE DEPTHS

What happens once we have fallen asleep? The obvious prediction is that we continue to move down-spectrum, to "sink into ourselves," until a turning point where we reach bottom and start upward again. We might go further, guessing the meaning of "sink into ourselves."

First we move from outer to inner reality. Then what? How would we sink deeper? By traveling further back in time.

It is hard to gather data about the subjective realities of sleeping and dreaming. But sleep-lab studies over the last half century have turned up useful information. In a 1965 study titled "Temporal Reference of Manifest Dream Content," Paul Verdone reported the following trends:

> During the first 3.5 hours of the sleep period dream reports referred to elements encountered in reality *in the last week*; during the next 4 hours, temporal references *moved back in time toward more remote events* . . . until approximately 7.5 hours of the sleep period had elapsed, when a *reversal of this trend occurred* toward more recent temporal reference.[13]

This is only one study, more a suggestive anecdote than hard data. But it *is* suggestive. Perhaps we do continue to move down-spectrum and "sink into ourselves" as we sleep, until we reach a turning point toward morning. Still more interesting, to *sink into yourself* once you have passed the sleep threshold, once inner reality wholly dominates outer, might well mean to sink into the past.

I have claimed that we are chronicle composers observing our own lives, living largely in retrospect. The metaphor shows us the importance of the growing vividness of experience as we move down-spectrum. In sleep-onset thought and dreams, *the past returns*. There we stand again in the presence of people we love who are gone—or people and places who (loved, hated, neither) were central to our childhoods. We live through these experiences, yet they never *happened* to us, because we do not remember them.

Still, repeated experiences of this kind—and they happen to us nearly every night—can change our selves in other ways. They can subtly bend our personalities by direct pressure or leave us with the haunting sense that something important has taken place just over the horizon.

A Summary Rule of Thumb

In summary: At the top of the spectrum, traffic flows mainly from consciousness to memory, to be laid down as memories for recollection in the future. At the bottom, that flow is reversed. Traffic goes mainly from memory to consciousness; memory supplies the sequences of recollections on which dreams are based.

Of course, the up-spectrum mind can and does recollect. But (again) it makes a disciplined, often abstract use of memory. And as we'll see, dreaming is first and foremost *recollection*—not creating, but reexperiencing, memories. The flow is so marked that recollections that would ordinarily have been turned back at the border of consciousness (as too disruptive or too painful) are allowed to push their way in. Those disruptive or painful memories find it easier to slip into consciousness in the lower (less watchful, less disciplined, less self-controlled) half of the spectrum. But they don't slip *consistently* into consciousness until we dream. Dreaming *is* remembering.

Spectrum Rule of Thumb: Up-spectrum, consciousness feeds memory. Down-spectrum, memory feeds consciousness.

This rule of thumb has a corollary, a related heuristic or useful approximation. Roughly halfway down-spectrum, the mental tide turns; the net flow changes direction. Henceforth, the flow of stuff *from memory into conscious mind* increasingly dominates flow in the other direction. And this change of direction has an important consequence.

Spectrum Rule of Thumb: Below the spectrum's midline, *thought is increasingly out of control.*

Fine, but it is out of *whose* control? Conscious mind notices what is happening—our rational, reflective self. Conscious mind is always on the job, even during the wildest, most flagrant dreams. But its part has steadily faded. The mental time and energy it commands has grown smaller and smaller.

Once we sink below the hallucination line, however, it is no longer *thought* that is out of control. Memory is no longer creating and pushing forward *thoughts*; it is creating reality. In dreaming, it is not thought but *reality* that is out of control.

AXES, IN SUM

In conjunction with the three axes I've described in this chapter—the logic-versus-narrative axis, the emotions axis, and the axis of outer to inner consciousness—a series of separate, related ideas arise:

1. The theme circle, or a theme-circling narrative, as a well-structured alternative to the plot-developing narrative.
2. States of action versus states of being as the two major kinds of mental states, and the associated idea of overconsciousness, of pure being, with its connection to poor memory function and dreaming.
3. The routine accessibility of past time and long-ago experience as part of sleep-onset thought and dreaming—with the rigid house rule that you must take nothing (or almost nothing) away from these encounters. But such encounters are, natu-

rally, overwhelming. And if we are overwhelmed, and forget ourselves or lose ourselves in the experience, it's not surprising that insufficient mental resources are left over to create solid memories.

We could draw three separate spectra instead of the one "master spectrum" I've described. One reaches from *reasoning* to *storytelling*. It might be labeled "How we make sense." One stretches from *acting* to *being*, labeled "Energy for doing versus existing." The last axis stretches from *outer field dominant* to *inner field dominant* and is labeled "Living now versus living then." They are all true models of the mind. We can and should reduce each separate axis to fundamentals. Any intellectually, scientifically serious discussion will search for basic elements that are as simple as possible.

Unfortunately, nothing will change the fact that the master spectrum itself consists of all three (relatively) simple elements overlaid, acting together. The finished product is, obviously, more complicated than its components.

PERSONALITY

I have looked at one transformation from several related viewpoints. We can also see the spectrum's reality (if not the actual *transformation*, the rising tide) in different personalities and personal styles.

Casual, unconsidered observations are often the most revealing. John von Neumann, the Hungarian Jew who emigrated to the United States in 1930, is often called the greatest mathematician of the twentieth century. The eminent physicist Eugene Wigner said, "Whenever I talked with von Neumann, I always had the impression that only he was fully awake."[14] The top of the spectrum, after all, is where we find logical (and, *a fortiori*, mathematical) thinking—*and* wide-awakeness. A first-rate mathematical genius soars higher in his logical thought than nearly anyone else. In the spectral (or ultraspectral) region of "exceptional logical capacity," our model predicts that von Neumann would also, simultaneously, be in the region of "exceptional wide-awakeness." This is a fine prediction, but what does it mean? Can a person be *wider* than wide-awake?

Wigner tells us yes. That was exactly von Neumann.

This sort of anecdote is not *data*—nothing of the sort. It is one small sandbag of plausibility, to help support a new approach on the rough, windswept plateau of "This is not orthodox!" "This is not the way we do things!" But that is exactly my task: only to make a new approach *plausible*.

At the other end is a subtler question: How low can we go, yet remain awake and aware? Napoléon was a different kind of genius. He knew warfare, politics, symbolism, big pictures. He knew power and how to wield it. His favorite topic was literature. Above all, he was a genius at handling people. "He knew," writes the celebrated memoirist Madame de Rémusat, "how to attract and deflect attention, to excite approbation on one side or the other, to upset or reassure whomever he needed to, to make use of surprise and of hope" (*Mémoires de Madame de Rémusat*). In his classic biography, Jacques Bainville quotes Napoléon as a young officer: I do "a thousand projects every night as I fall asleep" (*Napoléon*). His brilliance is clearest not in his wide-awakeness, but in the depth and breadth and clarity of his almost-asleepness, when he wanders the mind meadows of his particular, spacious-at-the-bottom mental spectrum.

Just as von Neumann seemed to push the spectrum upward far beyond its normal bounds, Napoléon seemed to push it downward, in the sense of putting far more space than usual between the start of drowsy free association and the arrival of sleep. Instead of falling straight to sleep, he could make use of the strange mind states of the lower spectrum. I'm not speaking of an abnormal ability to be exhausted yet resist sleep. I mean an abnormal *spectrum* in which the need for sleep isn't felt until farther than usual in the down-spectrum trip. Or, in which the mental energy released by the satisfaction of "doing a thousand projects" in bed with your eyes closed keeps a mind afloat and awake that would otherwise have long since sunk into sleep.

Von Neumann extended his spectrum into the upper reaches. Napoléon's extension, or lengthening, of the spectrum occurred near the bottom. Thus we see roughly the same phenomenon leading to very different consequences. Proust's alter ego speaks of his "natural inclination to daydream" (*À la recherche du temps perdu* [*In Search of Lost Time*]). Something like Napoléon.

Jane Austen's Emma tells us that "a linguist, a grammarian," "even a mathematician" is very different from her; *she* is "an imaginist." Napoléon was too—although Jane Austen, English patriot that she was, would have hated to say so.

"Her brother, she knew, was born to feel" (Cynthia Ozick, *Foreign Bodies*). Some people are.

One way of recognizing high-spectrum personalities is to observe how they handle many simultaneous inputs. With their gift for focus, such people can often concentrate on what's before them and tune the rest of the world out. You can speak to them and they will not hear—whether they're hard at work, or just reading a newspaper. Low-focus personalities cannot do this trick; they *cannot* tune out music, TV, or conversation in the background. The problem isn't noise; it's information. Noise in itself is no distraction, or not much, but *meaningful* noise is impossible for low-focus personalities not to hear. On the other hand, low-focus personalities are more likely than up-spectrum types to be tactful and knowing with other people. Because they cannot shut down their mental input ports, they are doomed to gather, constantly, the small hints and little clues about human nature that add up to detailed knowledge of human beings. However brilliant high-focus types might be, this sort of knowledge usually eludes them.

Keats had a different kind of low-spectrum genius. He was able to reach a state of perfect quiet watching, of *near*-pure experience where the mind, perfectly dilate, floods with being. The average person is nearly asleep at the point of reaching such a state. But Keats was able to be (just *be*), yet remain awake and aware. And he had exactly the phrase to describe this state: "wide quietness" in "Ode to Psyche," "visions wide" in "The Eve of St. Agnes."

The greatest of all expressions of the dilate, ultra-low-focus mind is Keats's "Ode to a Nightingale": "I cannot see what flowers are at my feet, / Nor what soft incense hangs upon the boughs." In the dark he is a passive absorber. The sounds, fragrance, and feel of the night take the place of dreams. "Darkling I listen; and, for many a time / I have been half in love with easeful Death." Preternatural receptivity with preternatural passivity—so *passive yet awake* that it feels, perhaps, like death. To be "laid asleep / in body, and become a living soul" (Wordsworth, "Tintern Abbey"); medieval painters show death as the living soul, in the form of an infant, soaring free of the body.

The nightingale departs.

> *... thy plaintive anthem fades ...*
> *Was it a vision, or a waking dream?*
> *Fled is that music:—Do I wake or sleep?*

Ponder the expression on Baldassare Castiglione's face in Raphael's great portrait (perhaps the greatest ever painted) in the Louvre. You see a man with his mind dilate, taking it all in, drinking life in deep, quiet drafts. Where others find a teaspoonful of life, he finds endless depths.

Everyone has a distinct cognitive personality or thought style—

a cognitive *gait*—which can change or mature or evolve. Hölderlin writes memorably, in a short poem called "Ehmals und jetzt" ("Then and Now"):

In younger days I rejoiced every morning,
wept every evening; now that I am older,
I start my day doubtful—and yet
Its end, for me, is sacred and serene.

The stronger your grasp of every second of your life, including the paradoxical experience at the bottom of the spectrum, the stronger you are.

THE BIGGER PICTURE

Only subjectivists and subjectivism can elucidate the deep problems of the spectrum, or at least inch forward in that direction—because only subjectivists can *see* them. Martin Heidegger was a terribly weak man but a strong philosopher, and he is never more impressive in his superb astonishment at sheer being (why *am* I?) than when he points to one of van Gogh's paintings of worn peasant's shoes and tells us, with a shrug, anyway—*he* understands. Van Gogh's painting throbs, pulses, practically pants out loud with the pain and elation of this monumental *is*. The shoes (amazingly!) *are*, and van Gogh has painted that awe-inspiring fact. Others, easily a dozen among the greatest Western painters and sculptors, have done likewise. They have *painted* overconsciousness, consciousness burn—an important element in Western art that we have never properly acknowledged.

Van Gogh's paintings of shoes dare us to ask, What does this painting show?—and to answer, *Nothing!* Van Gogh was not indifferent to his subjects, but the way he carved and sculpted his images out of paint counted much more. All his mature paintings and drawings struggle to show sheer *being*. It is no accident that Cézanne should have been obsessed with the same thing in his own way. What matters is not what the painting is about (its "intentional content," metaphorically speaking), but how the painting *is*; the way it has chosen to *be*—striving mightily to make us see the actual fact of being.

But there is also a less obvious, more powerful way in which these paintings shed intentionality and reached for absolute concreteness, pure being—just as intellectual life, especially physics and mathematics (which set the tone), were mastering ever-higher peaks of abstraction. We see painters pouring terrific, transcendent energy into the struggle to show being. Just being in itself, *being there*, right *there* before your eyes (*Dasein* to Heidegger—although he uses the simple phrase to refer to essential human-ness). "Your life's a *miracle!*" says Edgar to Gloucester in *King Lear*, after the miserable, blinded Gloucester has tried and failed to kill himself. *Every last thing is a miracle*, reply the painters.

We have not even begun to grasp this bottom rung of the spectrum—just *being*; but look, here is *something* where there might have been nothing! Astounding. (This something where there might have been nothing is the basis of Leibniz's famous, controversial, "cosmological proof" for the existence of God).[15] Above all, the *human* being is a miracle. The mind has followed an inconceivably daring and beautiful path from raw nothing to Percy Bysshe Shelley writing: "I am the eye with which the Universe / Beholds itself, and knows it is divine" ("Hymn of Apollo").

Thus, Cézanne paints Ambrose Vollard, van Gogh paints Dr. Paul Gachet; Picasso paints Gertrude Stein; Soutine paints anyone; Alberto Giacometti paints his brother Diego. Each of these images is a monumental struggle to depict what is undepictable, the model's sheer existence—in a universe that is mostly full of nothing.

But in the last analysis (caution!), these paintings are nonetheless *about* nothing—*not* about being or about anything else. They are not *about*; they just *are*. They don't argue or demonstrate. They are, and that's that. It is what they are as objects with paint on them, not what they *show* in imaginary picture space, that makes them powerful. They *are* the artist's struggle with the fact of being, with the *something* that might have been nothing—yet, somehow, isn't.

Thus we saw in the twentieth century the mainstream scientists climbing higher, rung by rung up the spectrum, like high-wire artists stepping up calmly into a stratosphere that is so far overhead, we can't even see it. They are confronted (so to speak—in an inverse sense) by a vibrant artistic community going just the other way— matter and antimatter. The artists were deliberately *descending* the spectrum rung by rung as far as they could go toward the absolute bottom—not toward pure abstraction, but the exact opposite, toward pure concreteness, where sheer being is a miracle that needs grasping as much as or *more* than it needs scientific explanation. It has been explained but has yet to be plainly *grasped*. (Recall Dilman: "What he needs is a *clarification* of the concept of consciousness, instead of an *explanation of it along scientific lines*.")[16] Thus the spectrum becomes a metaphor for the most dramatic split between science and art we have ever known.

My task in this book is to assemble the fundamental facts about subjective reality, show you what they are, and try to convince you that they amount to a picture of the mind in motion, to the spectrum of consciousness.

Three

Every Day

Every day we move down a continuum of physical states from rested and relatively wide-awake to tired and needing sleep and, beyond, into sleep and dreams. That spectrum of physical states has a corresponding spectrum of mind, of qualities of consciousness. Up-spectrum, we are focused on the outer world, on abstraction and reason. As we move gradually from alert to relaxed to drained to drowsy, withdrawn, falling asleep, our mental focus keeps declining. Eventually, our capacities to reason and reflect burn down like candles to virtual silence—as we hold tight on our tumbling, plunging, twisting-and-turning coaster ride through the haunted house of dreams.

But we don't stop sinking—one of the mind's many surprises. As we sink from outer to inner, objective to subjective, awake to asleep, we also sink right out of the present into the past. Emotion draws us irresistibly. Love story, horror story—it doesn't much matter; it is the *intensity* that draws us. Rational intelligence fends off emotion-ridden memories, which disrupt straight thinking. But rationality has grown tired and fallen asleep, and emotional memories are free to flood consciousness. Our *most* emotional memories are usually childhood memories, often early-childhood memories. Since we no longer inhabit objective reality, each of those memories is an alternate reality. They draw us out of *now* altogether, into the past. So we sink out of objectivity, rationality, awakeness—and plunge on into the past.

In Chapter 2 I traced the transformation from three viewpoints. It's time to braid them together. Up-spectrum, we make sense of the world by reasoning, abstraction, and informal logic, but that gradually changes, and storytelling becomes our chosen method. Meanwhile the *acting* of rational reflection gradually takes on emotional and sensory color and becomes *being*: how you *are* dominates what you *do*.

Emotion grows increasingly prominent as reflective thinking fades and the brightness of memories grows—and, by *not* creating memories, we *unmake* our own experience as it happens.

To the human mind, the world of pure being is just as important as the up-spectrum world of pure thought—although many researchers would rather ignore it.

The mind moving through the spectrum is a complex and beautiful mechanism. It moves from active to passive, present to past—as the themes that define our lives emerge, circled by stories. But the substance of these themes is merely hinted at by a Stonehenge of signposts, all pointing to the same pregnant silence in the middle.

My discussion of the tide in this chapter, and of the native landscape that is swept by the tide in the next, will lay the groundwork for a more careful exploration of the spectrum and its consequences.

From the Top

Sometimes we wake like popped balloons, because an unpleasant stimulus punctures our sleep. Sometimes we are able to sleep until we have slept enough, and we awaken as naturally as swimmers coming out of the water up a gentle slope of beach. Sometimes, as we come awake, we lapse back into light dreams—the kind that are hardest to distinguish from reality, that leave us unsure whether we *have* been asleep. Sometimes, at gentle awakenings, we keep handling the same material that occupied us during the last dream—continuing to spin a narrative based on absurd or at least slightly odd settings and characters, *although we are awake*. Usually we put these dream topics aside unthinkingly and move on. The material itself is as fugitive as any dream. It is forgotten in a moment. But occasionally, we notice with puzzlement what we have just had in mind.

On these unusual occasions we grasp, by reflecting afterward, the basic character of dream thought. Its job is to accept material from recollections, the imagination, and sometimes perception, and *work it up into a narrative*, into something we might call "counter-reality." Up-spectrum, the mind's job is to build a reasonable path from premise to goal. The low-focus, dreaming mind is the ultimate improvisational theater.

Why do we dream? For the same reason we think. Freud famously believed that *a dream is the fulfillment of a wish*. He meant that a dream presents a story—often disguised, distorted, or incomplete—of accomplishing something we want. Jonathan Lear emphasizes that this Freudian "wish" is a creature of the dream world, intended for *fulfillment* in the dream world. It is not a mere ordinary wish that cannot be satisfied because we are asleep. On its own terms, it *is* satisfied.

Freud's answer was, in fact, broader and deeper than the one statement. Yet it was only partly correct. On the other hand, the idea many people have that Freud was "just wrong about dreams" is like arguing that Euclid was just wrong about geometry. Here is what a popular psychology textbook says on the topic:

> Sigmund Freud proposed that dreams were disguised outlets
> for the inner conflicts of our unconscious. . . . But given its lack

of scientific support, the Freudian view has been rejected by most contemporary sleep researchers.[1]

If you want a good laugh, read some modern psychoanalytic case histories—for example Morton Reiser's or Stephen Grosz's—and then try saying "dreams are *not* disguised outlets for inner conflicts" with a straight face. But the best way to learn that Freud was right in this instance is to observe and think about your own mental life and your own dreams. Jonathan Lear again: "Until a person is able to fill up those [Freudian] concepts with their manifestations in his own life, his understanding of those concepts will be hollow."[2] Lear believes that, in modern culture, it's not Freud and his depth psychology of the individual that has become obsolete; the *individual* is becoming obsolete.[3]

Notice that in answering "why dream?" with "to fulfill a wish," Freud was doing something fascinating that is often overlooked. He was saying, in effect, that *we dream at night for the same reason we think during the day.* The goals of dreaming are just the goals of thinking. Freud knew well that dreaming is a form of thought. ("At bottom," recall, "dreams are nothing more than a particular form of thinking.")[4] Waking thought centers on planning ways to get what we need, or do what we want—in other words, to fulfill our wishes. Waking thought also solves any problems we wish (and are able) to solve, reacts to the physical environment, and so on. But on Freud's reading, the overlap between waking thought and dream thought is much larger than the difference.

Plainly, Freud's view is correct. Yet it is also fair to say that the main function of dreaming is not wish fulfillment. Its main function is simpler and more obvious. Freud himself might even have agreed to it. I will examine this function at the end of the chapter.

Waking

I have been speaking of dreams because we start our days by waking, often out of dreams. But let's re-join the awakening process itself. Different people and occasions call for different stimuli to cause awakening. Ordinarily, it takes us some minutes to come fully awake. (At breakfast: "I didn't hear your chair I was so lost in my dreams, didn't realize you had stopped talking" [Sean O'Reilly, *Watermark*].) Sometimes we use the same word, "sleepy," to describe our states before and right after sleep. We are in the same state both times: far down-spectrum, occupied by what's inside our minds, neither able nor willing to focus on the outer world. We must re-aim consciousness from the inner to the outer field—a portentous operation. "If, following awakening," dream-lab researchers write, "we do not abruptly abandon the dream world, but try to retain contact with it, there is a greater likelihood to recall a dream."[5] If we deliberately "keep looking" at the *inner* field of consciousness, the dream world, we are more likely to remember what went on there.

"He's scarce awake, let him alone," Cordelia is counseled in *King Lear* when she is desperate to talk to her father but he has just awakened and needs time to come to himself. When we are awake but not *quite*, we have the sensation, sometimes, of looking from the bottom of a clear, shallow lagoon at the wide-awake world above the surface; of uncoiling gracefully yet uncertainly upward, like a sea plant. "Early next morning, in the nameless space between sleeping and waking, he has a dream or a vision" (Coetzee, *Childhood of Jesus*). So there *is* a "space" within which one can dip back into dreaming—*light* dreaming; often your actual, real-world surroundings are part of the dream.

To come awake means to travel up-spectrum, often with a jolt—from sleep to waking, and past the hallucination threshold. We journey from taking inner consciousness as basic reality to treating outer, real-world consciousness as basic. Perhaps physiology makes it simpler for us to make the up-spectrum trip than to move down-spectrum. Sheer survival might depend on our snapping awake quickly. It rarely depends on our falling right to sleep. Sometimes, in living things and machinery, the same mechanism that makes it easy to go one way makes it hard to go the other. Is this true of spectrum physiology? We don't know.

Still, waking is (to say the least) not always easy. "Like ice forming in water, memories begin at last to coagulate: who he is, where he is" (Coetzee, *The Master of Petersburg*). Saul Bellow's self-absorbed narrator in *Humboldt's Gift* wants to know "why waking was so convulsive." "Her spine felt drilled through; her brain still swarmed with fearsome dream-shreds retreating into oblivion. She had slept hard and wickedly. Her dreams were rife with treacheries" (Cynthia Ozick, *Foreign Bodies*). Her "dream-shreds *retreating*": a powerful phrase, just right. We feel dreams pulling away like living things, huddling away from us as we reach out helplessly.

We do know that you cannot force your way down-spectrum, however much you'd like to. We also know that, ordinarily, we spend the whole day and perhaps part of the night traveling down-spectrum, but it takes far less time to come fully awake and move straight up to the top.

Having come awake and reached the spectrum's top, we oscillate down and up again during the day. We descend in a series of swallow swoops, down and partway up, further down and partway back up again. The swallow swoops continue in a different sense while we sleep: we dip into deep sleep without much dreaming, ascend to lighter, more brain-active periods (REM sleep) where most of our dreaming occurs, and then repeat. We complete four or five such swoops on an average night.

Much research has focused on circadian cycles, in search of normal or typical patterns of energy and sleepiness. They are hard to pin down; people's inner clocks differ greatly. About half the population seems to consist of distinctly "morning" or "night" people. Beyond that, most people's mental energy or focus level seems to rise as they gradually come awake until a midmorning maximum; then, focus declines until midafternoon. After this midafternoon drowsiness point, energy and focus level drift back up again until early or midevening, when the last downward stretch begins. Thus, there seems to be a natural evening peak, presumably more pronounced in "night people" and those who prefer to concentrate on their work toward the end and not the start of the day. In this view, focus rises, falls, rises once more, and then falls continuously into sleep.

Office Hours

If the spectrum does bottom out for many people in midafternoon, that would be around 3:00 p.m. "At three in the afternoon—the hour when, all over the world, the literary stewpot boils over, when gossip . . . is most untamed and swarming" (Cynthia Ozick, *Messiah of Stockholm*). If gossip is the opposite of focused work, if gossip is the bearing of tales or, in other words, storytelling, we have just what we would expect if all sorts of people were indeed to reach a spectrum low point in midafternoon.

No workday schedule could possibly be ideal for more than a small fraction of workers. Discovering the details of one's own spectrum—which times (ideally) to do what sort of work, when to rest—is one of the great projects of young adulthood. But many people never really get to do it.

Many office workers would probably do better under a split schedule than a standard workday: 8:00 a.m.–1:00 p.m. (say) would be the quiet, hard-work period, with minimal interruptions. No meetings, no phone or video calls; no electronic messages until noon. (Obviously it can't work in every case. But for some people, in some jobs, it can.) Then there would be a break until, say, 6:00 p.m., and a second three-hour work period for more "reflective" tasks, meetings, and communication. Then everyone goes home.

Would people be happier and more productive this way? Siesta countries already have schedules something like this, and it doesn't seem to do *them* much good. But a *disciplined* first segment is essential. Many office workers would get more done in a disciplined, early-in-the-day five-hour period, with strictly limited interruptions, than they do in today's eight standard hours. If they worked well, you would send them home for the day at 1:00 p.m.

Unfortunately, these are 1920s-style experiments. We aren't likely to see any nowadays. But they're interesting to contemplate.

These ideas—the split day with two work periods of different types—are all based on the daily circadian cycle, which is well known. But Eric Klinger sees another cycle governing daydreams.[6] Daydreaming occurs in roughly ninety-minute cycles throughout the day, and as each period nears its close, we grow more likely to be distracted and to daydream. Of course, there is always a limit to any period of close concentration, whether we are up-spectrum and fairly fresh or down-spectrum and working doggedly; ninety minutes (also the length of the typical sleep cycle) seems a convincing average value. Yet we are clearly far more distractible, far more likely to be preoccupied and get lost in a daydream, when energy and focus are low, down-spectrum, than at high energy and focus near the spectrum's top.

Natural cycles are important to any inherently cyclic, tidal process like the rise and fall of mental focus. But we must treat such news cautiously. The variation between one personality and another, one job and another—between, for that matter, Monday and Friday—might easily swamp any underlying pattern. It's safe to say that the basic trend over our waking hours is from top to bottom of the spectrum.

Travels across the Spectrum

"She was in that highly-wrought state when the reasoning powers act with great rapidity; the state a man is in before a battle or a struggle, in danger, and at the decisive moments of his life." Thus Tolstoy wrote in *Anna Karenina* (Garnett translation), describing the most focused place on the spectrum, the very top. "The state a man is in before a battle or a struggle" suggests that we can almost always force ourselves up-spectrum. But we can never force ourselves down.

We would often like to send ourselves down-spectrum, in order to relax or sleep. Alcohol and other pleasure drugs exist mainly to break the iron grip of high focus and let us sink at least a few rungs down-spectrum. Those who can master the discipline of meditation move down that way. But it takes practice and does not come easily. Children have been known to decide they want to sleep and then simply relax their bodies and (almost immediately) fall asleep. But this is a hard trick for adults to pull off.

No matter the age, what is perhaps most important is that movement over the spectrum is not just *mental* movement—not just a decline in mental focus, self-control, and acuity, and a rise in other ways of thinking, remembering, and being conscious. It is also a *physical* change, a change (as we move down) in muscle tone and tension. (In some sense, downward movement might also be described as a change in "mental tone.") The more we learn about the spectrum, the more we discover how close—and how often overlooked—are the connections between states of body and of mind. We need a brain *and* a body to make a mind.

This is one of John Donne's most important themes. He is a

devout Christian and preacher and understands body and soul as the components of mind; and of the two, the body is decisive. "I say again, that the body makes the mind" ("that the gifts of the body are better than those of the mind" ["Paradox VI"]).

We see the intimacy of body and mind in countless ways; happiness, for example, is the greatest of all energy boosters. "By the mass, 'tis morning;" says Iago the villain of *Othello*, after a happy night spent scheming and ruining people. "Pleasure and action make the hours seem short." It is not surprising that sadness should make us listless and tired and draw us down-spectrum. "Commend me to thy lady," says the friar to the nurse in *Romeo and Juliet*, "And bid her hasten all the house to bed, / Which heavy sorrow makes them apt unto."

And just as happiness increases mental energy, sheer mental energy (other things being equal) makes us happy.

Emotions grow gradually more important as we move down-spectrum, in several ways. As we tire, naturally we lose control. Asleep, we have virtually none over body or mind. Gradual movement down-spectrum means a gradual loss of self-control—which affects the degree to which we are "emotional" in the sense of letting our emotions show. If a friend makes you angry, at the focused start of the day you are likely (other things being equal) to get on with your business rather than berate him. When you are tired after a long day, you are more likely to light into him. In between, you're angry and you say so, but you rein yourself in. Big scenes tend to come lower on the spectrum.

By one definition, you have become more emotional by moving down-spectrum and losing self-control. Internally, you might have been just as angry in the morning as in the evening. But expressing our anger often heightens it; showing our anger can increase our internal, felt anger.

Getting drunk reproduces some aspects of a swift down-spectrum tumble. Drunk, most people seem to be more emotional externally *and* internally. Often they seem to feel more acutely, reinforcing our guess that self-control affects not only emotions expressed but emotions experienced.

We must evaluate the entire spectrum, and that includes foreign relations—how different spectrum points relate to the world at large. Ours is a high-focus world. I will write in detail about the power of low-focus thought, because the low end of the spectrum is the less understood by far. But let's not kid ourselves. We don't think much about it, but our world is built by high-focus thinking—built not just by science or formal reasoning, but by the long progress of practical engineering since antiquity.

Practical engineering is reasoning with your hands. It is up-spectrum work. We live in a world created by the technology of ceramics, masonry, glass, metal, textiles, mills and clocks, and hand tools; measuring and navigating tools; presses and pumps; and paper makers, roads, wheels, springs, blades, and weapons of every sort. Our world was created by wire-making, gear-making, and gear-train-making machinery, and engines of all kinds; by power generators, amplifiers, transformers, power tools, machine tools, lighting and heating and cooling and refining, distilling, extracting, synthesizing. The amount of intellectual progress we never waste a thought on is staggering.

Two Fields of Consciousness

Let's return to the mind itself—not forgetting its all-but-incredible accomplishments. There are two fields of consciousness, outer and inner. Outer means perceptions, of the external world and our own bodies. Inner means recollections, and ideas we concoct. We can perceive and examine (metaphorically) things we have recollected or invented—*inner* things, pink elephants and pale blue peaches and flying hamburgers—just as we perceive and examine the outer world.

For many people, visual thinking—including the invention and manipulation of abstract images—is crucially important. But visual thinking is poorly understood.[7] For many people (few of whom seem to be philosophers, unfortunately), *pictures* are the language of thought. "Dreaming is continuous with our waking *reflective ability to think in images*," writes Foulkes.[8] But most philosophers, and perhaps most psychologists too, are more comfortable with language than with pictures as a thought medium.

So there are two fields of consciousness, each able to show us facts in many forms. And whenever we are conscious, we are conscious of both fields. But at the top of the spectrum, the outer field dominates. At the top, "reality" means outer, physical reality. At the bottom, or near the bottom (at the absolute bottom we are unconscious), the inner field dominates. When we dream, "reality" means inner, hallucinated reality. Of course, we don't switch over instantly from dominance by the outer field to dominance by the inner field. The daydreams and fantasies of the lower spectrum represent inner reality growing continuously more attention-grabbing and vivid.

Thus, one way to understand the mind's transformation as it moves from top to bottom is that it gradually pulls itself inside out. The two fields of consciousness, outer and inner, might be the outer and inner surfaces of a sphere. The sphere has a hole at one pole and is easily reshaped. Up-spectrum, the outer field dominates the inner. It makes the outer surface. The inner field, the inner surface of this sphere, is farther away and less vivid. As we move down-spectrum, the sphere is gradually flattened: it becomes a disk—outer field on one side, inner on the other. As we continue down-spectrum, the disk is worked into a sphere again, this time with *inner* consciousness on the outside. Consciousness has been pulled inside out.

But our move down-spectrum is no simple transfer of focus from outer to inner. The mind *does* something and simultaneously *is* something. As we move down-spectrum, we are increasingly aware of our state of being—what we sense (meaning the sensations themselves) and how we feel.

The Middle Regions

The middle parts of the spectrum are often hardest to notice and speak about precisely. They are the moments that are merely *normal*. Thus Jane Austen's Emma, as she watches the minor, mundane doings on her local Main Street: "A mind lively and at ease, can do with seeing nothing, and can see nothing that does not answer" (*Emma*). Anything she sees, however minor, will answer the needs of her lively, easy mind. She watches—is fully aware of—the ordinary doings on the uncrowded street. That is outer-field consciousness, and it is not *quite* enough to entertain her. Instead of turning away and seeking something else, her lively mind "can do with seeing nothing." Her own thoughts are interesting enough to piece out the entertainment. Inner-field consciousness is also a part of her world, as it is of everyone's—but some people enjoy it more than others. Emma, at this point, is somewhere in the middle of her spectrum. And the scene takes place, appropriately, around the middle of her day.

If it had been later and Emma had been further down-spectrum, she might just have tuned out the outer world and daydreamed. She might have turned from outer- to inner-field consciousness—still vaguely aware of the outer world, but preoccupied. "The girl had been quiet and a little withdrawn, since she had seen Alvarito. Something was going on in her mind. . . . Momentarily, she was not with them" (Hemingway, *Across the River and into the Trees*). This small incident of temporary distraction happens in the evening, in the dark of a gondola—as Emma's pausing between outer and inner worlds happens at midday. The authors have no thought of the spectrum. But certain incidents feel right in certain places. The spectrum emerges in this way implicitly, as a matter of instinct. And this sort of instinct is part of the large gap between ordinary writers and great ones.

Sometimes we choose to daydream. ("I drew up a deck chair in a sunny spot, closed my eyes, and indulged in a little day-dreaming" [Coetzee, *Summertime*].) Other times, daydreams choose us. The psychologist and pioneer daydream researcher Jerome Singer finds that most daydreams are *involuntary*.[9] They take control of our minds uninvited. "Willy-nilly, he finds himself slipping into daydreams" (Coetzee, *Childhood of Jesus*). "He was haunted by daydreams and such strange daydreams" (Dostoyevsky, *Crime and Punishment*).

A daydream often leads us away from the mental path we started on. We *lose control* of our thoughts as we descend the spectrum— lose *conscious* control. We cede control to the unconscious parts of mind. ("When you approached her," writes André Gide of a day-dreamer, "her eyes would not turn from their reverie to look at you" [*La porte étroite* (*The Narrow Gate*)].)

Thus, daydreaming increasingly represents not a pleasant diversion but a loss of control as we move down-spectrum. Of course, not all daydreams are pleasant. Singer's studies showed, too, much "negative emotion" in daydreaming—"anxiety or anger or guilt, in addition to joy."[10] As we gradually lose control over our minds, unpleasant thoughts are more likely to barge into consciousness.

The Liberation of Emotions

The stormier climate of daydreams as we move down-spectrum fits perfectly with the liberation of emotions. We are apt to be more emotional when we are down-spectrum and tired than when we are up-spectrum and alert. "It is awfully easy to be hard-boiled about everything in the daytime, but at night it is another thing" (Hemingway, *The Sun Also Rises*). Emotion spreads gradually into the mental atmosphere as we move down spectrum, perfuming and changing it. We don't suddenly become emotional when we cross a particular line on the way down. Upon entering a dark room, your eyes adjust gradually; the objects around you begin to emerge. And similarly, as you move down-spectrum, consciousness changes gradually from unemotional black-and-white to a million-colored patchwork—and nearly all the colors are subtle and quiet, mere color-washed grays. Then, slowly, your thoughts become a heath in pale winter sun, silvery green and moss brown dotted with a million tiny white, lavender, and rose florets. And finally, the brighter stuff starts to appear. Crabapple. Azalea. Emotions can be loud and hot.

One subject of a study (in this respect fairly typical) sometimes uses harsh words coolly in the morning, and always regrets them in the afternoon. His rare outbursts of real anger or temper, on the other hand, come in the evenings—always accompanied by the idea that he will be more rational in the morning. In adolescence and college, he would often spend much of the afternoon practicing the piano, but if he played in the evening, he played only for pleasure or he improvised and wrote music. He associates early-morning classes during his high school and college days with a sort of numb stiffness. Classes were more relaxed and fun as the day wore on. And so on. Many small things.

Of course he was capable of feeling emotion in the upper spectrum. He remembers long-ago *and* modern times when he was miserable in the early mornings because of the agenda for the day. But on those mornings, he would seldom say much about his unhappiness. The emotion stayed inside, bound and gagged, and he would set out without hesitating, passively miserable but uncomplaining. In lower-spectrum periods, he complained plenty.

Emotions are fundamental to the mind and thought in many ways. One, not appreciated sufficiently but important, is the fact that an emotion is a meaning-independent summary. Because many complex facts of a scene, person, or recollection can be captured by one subtle, nameless emotion, emotion is crucial to remembering— to grabbing out of memory the stored information we need in order to live. *And* because emotion is a meaning-independent summary— two radically different things can give rise to the same emotion—it can bring about the invention of new analogies.

I'll return later to the peculiar information-processing power of emotions. Emotions have other crucial properties too; they set thought in motion by determining our needs and wants, our fears, anxieties, and hatreds. Emotion runs the mind. Thus the great British empiricist philosopher David Hume: "Reason is, and ought only to be, the slave of the passions."[11] And many others. A Philip Roth character says, "If it weren't for sentiment, Zuckerman, one person would not pass another person a glass of water" (*The Prague Orgy*).

Another important property of emotion lies in creating personality. Those who are guarded or reserved are in a different personality class from the emotionally outgoing, from those who are happy (even eager) to show their emotions. The difference between the two might be in the character of their emotions, or their spectrum personalities, or both. Some people have more intense and vivid emotional responses. Some are especially at home in the lower spectrum, where emotion permeates the air like the smell of wine or pot or roses.

Your way of showing emotion is clearly one of the main determinants of your personality. This is one of Jane Austen's themes. In *Persuasion*, the heroine ponders a rejected candidate for her hand:

> Mr. Elliot was rational, discrete, polished,—but he was not open. There was never any burst of feeling, any warmth of indignation or delight, at the evil or good of others. This, to Anne, was a decided imperfection. . . . She prized the frank, the open-hearted, the eager character beyond all others.

One feels that Jane Austen did too, and that she was frank, open-hearted, and eager herself. Henry Austen wrote a famous sentence about his sister after her tragically premature death: "Her eloquent blood spoke through her modest cheek" (*Preface to Northanger Abbey and Persuasion*)—"her eloquent blood" meaning the richness, alertness, penetration of her emotions. Her color and complexion advertised her feelings, to those who knew. Reserve was the only flaw (but it was serious) in the personality of Emma's elegant, pretty, smart, and talented competitor Jane Fairfax. Henry Austen's words echo Donne's famous "Second Anniversary," in praise of a patron's child who died at fifteen:

> *Her pure and eloquent blood*
> *Spoke in her cheeks, and so distinctly wrought,*
> *That one might almost say, her body thought.*

"Her body thought" is no mere turn of phrase, but a fundamental conviction: "I say again that the body makes the mind." If philosophers of mind had considered the implications of Donne's doctrine—not the trivial fact that the brain is a physical thing, but the deep observation that interchange between body and mind is constant, and shapes our thoughts—they would have avoided many wrong turns, including (probably) computationalism.

"Exuberance is beauty," says William Blake (*Marriage of Heaven and Hell*). In some people, emotion lies deep and is rarely visible, or is so dilute as to be hard to make out. Maybe they dislike spending time in the lower spectrum, where emotion is let out to romp. More often (one guesses), their emotions are mostly unsaturated, inherently dilute; or they are set against disclosing emotions because of their upbringing, and have thrust them down deep. People like Jane Austen's *Persuasion* heroine Anne prefer those whose emotions are vivid, or nearer the surface: Anne knows what "open-hearted" people are feeling and thinking, and therefore has more chance to sympathize. ("Her own special talent: *meegevoel*, feeling-with" [Coetzee, *Summertime*].) To be interested in your fellow creatures means, furthermore, to react to them emotionally; we are interested in those who are interested in us. People we can't read, like books in languages we don't know, are frustrating and boring.

Being up-spectrum has personality implications in itself, because up-spectrum means mentally energetic. How do we understand, in spectrum terms, the tendency of happiness to make us talkative and outgoing, sadness to make us withdrawn? Back in *Persuasion*, Henrietta is preparing for her wedding:

> Henrietta was exactly in that state of . . . fresh-formed happiness, which made her full of regard and interest for every body she had ever liked before at all.

We have all seen such behavior. And, as I've mentioned, happiness increases mental energy, which sends us up-spectrum. Being up-spectrum promotes external versus internal focus, makes us attuned to the world around us, ready to act or react. She is happy, so she feels energetic; *that* pushes her up-spectrum and makes her outer-centered. Sadness lowers mental energy—which sends us down-spectrum. There we grow inner-centered, withdrawn, and even sadder.

But *does* happiness really make us mentally energetic? There's no abstract reason it should, but it does indeed; we find evidence in language. "She was jumping for joy." One *might* be jumping for sadness, but one rarely is. "He got into Columbia Law School; is he excited?" Is he eager and happy? One *might* say, "I hear you got fired. Are you excited?" But one rarely does. Happiness makes us energetic, which sends us up-spectrum, which makes us outgoing. That depressed persons should be withdrawn, speak little, and rarely go out (should be *inner*-centered) is part of the same broad principle that makes us dream when we sleep. The transition from high to low spectrum is a transition from a predominant outer field of consciousness to the inner field.

FANTASY

Let's continue down-spectrum, from daydreaming to fantasies. Fantasies are hard to distinguish from daydreams; in common use, the word "fantasies" often refers to daydreams with sexual topics. The psychology literature sometimes places fantasies farther down-spectrum than daydreams (without using the term "spectrum"). But this is a useful distinction: let's agree that fantasies are a step closer than daydreams to the strange twilight of sleep-onset thought, and to sleep itself. There is a lot we don't know about fantasies, and the fact of our relative ignorance is an important bit of information.

Fantasies that happen *to* us (rather than those we plan) are much like real dreams, except they are not hallucinated. One fascinating study found bizarre situations in fantasy nearly as often as in dreams. "Bizarre features are certainly not limited to dreams; they infiltrate the waking fantasy activity in manifold ways."[12] Let's look at a concrete example: selections from one typical fantasy gathered in a Swiss sleep lab. The subjects were lying in bed but were not asleep—eyes not closed.

I was inside my old school. . . . Then some young people came sliding down the banister, very fast, a whole lot of them. I then entered a workroom. . . . A miserable gray structure of pillars. It begins to shake and keeps crashing down . . . but my feet are underneath. And as I have my feet down there, they become like a spirit that emerges from a bottle. And then I get to be like smoke, on top.[13]

These descriptions were collected the way sleep reports are taken in sleep labs. Some people are surprised to learn that illogical, bizarre, and fantastic storytelling occurs in fantasies. Recall that these are *ordinary waking* fantasies; we are down-spectrum, our mental focus is diffuse, and we are not energetic, but we are fully awake in every sense. Why be surprised? And why use sleep-lab techniques to gather fantasies? Exactly because *we recall fantasies badly*, just as we do dreams. Fantasies that are vivid enough to pre-occupy us, keep our attention, keep us distracted, tend to be vivid (not as vivid as dreams, of course)—and thus tend, also, to create overconsciousness and bad remembering. We do tend to remember *dreams* on a regular basis, when we wake up in the middle of them. But we don't wake up from, *break away from*, fantasy in the same sense because we are awake already.

THE SPECTRUM'S CONTINUITY

The goal is to see the *continuity of the spectrum*. Dreamlike thinking doesn't emerge suddenly when we are asleep. Dreamlike thought is one way for the mind to be, and it grows increasingly frequent and pronounced as we move down-spectrum.

In the table shown below, the progression as we move down-spectrum—from out-of-control thought to sleep—is obvious.

	Out-of-control thought	Story-telling	Bizarre features	Hallucina-tions	Sleep
Thoughts wandering	+	O	O	O	O
Daydreaming	+	+	O	O	O
Fantasy	+	+	+	O	O
Sleep-onset thought	+	?	?	+	O
Dreaming	+	+	+	+	+

But nature in detail is rarely neat and tidy. Sleep-onset thought is anomalous in some ways: though each separate episode is like a small dream, the story is often implicit or submerged—and sometimes there is no story, merely a static scene.

Sleep-onset thought is sometimes bizarre, but less often (it appears) than dreams. It is a mixture of hallucinations and ordinary thought.

Losing Control over Our Own Thoughts

Before we descend the cliff face of consciousness and the spectrum any further, what does "loss of control" tell us about the structure of mind?

The great poet and philosopher Samuel Taylor Coleridge speaks of "master currents below the surface" taking over the mind, and of "flights of lawless speculation"—following no rules, no logical laws (*Biographia Literaria*).[14] We are all familiar with states of mind (or parts of the spectrum) where thoughts happen *to* us.

"C'est faux de dire: Je pense," wrote Rimbaud; "on devrait dire: *On me pense*" ("One should not say 'I think' but 'I am thought'" [*Letters*]).[15] In his soaring, gliding memoir *Speak, Memory*, Nabokov writes: "Just before falling asleep, I often become aware of a kind of one-sided conversation going on in an adjacent section of my mind, quite independently from the actual trend of my thoughts. It is a neutral, detached, anonymous voice, which I catch saying words of no importance to me whatever."

When we speak of losing control, we don't mean that a sorcerer or demon is taking over. Clearly, we mean that an autonomous part of the mind *itself*, outside our conscious control, with its own agenda (so to speak), is calling the shots. We mean that the *unconscious* mind is taking charge.

And what *is* the "unconscious mind"? Basically, it is *memory*.

Memory itself has two main components, for our purposes. (I focus here not on "procedural" memory for skills, but on "episodic" and "semantic" memory—where incidents of our lives are kept along with facts and ideas we have learned.) "Memory" means the stored information itself *and* procedures for storing and retrieving it.

The cars in a Manhattan parking garage *and* the attendants who park and fetch them are collectively the "49th Street Garage." The cars are the memories; they are stored and retrieved. The attendants are the storage and retrieval mechanism or procedures. One final function of memory is to maintain the collection in good order. This curatorial activity takes the form of "background processes" of which we are never aware. The most important processes allow us to learn by forgetting: by forgetting distinctions among many similar memories, by allowing them to blend together, to *melt* together into single (slightly heavyweight) memories, we make space, clear up clutter, *and* create abstract templates or schemata. I'll return to this topic in Chapter 5.

Ordinarily, all three aspects of memory, the two majors and a minor—the stockpile, the storing and fetching, the background processes—are unconscious. You have in your memory untold hordes of items of which you remain unconscious nearly all the time. You know the names of the Great Lakes and the music and lyrics of five hundred Beatles songs (or the equivalent for another band) but are rarely conscious of them. The memory processes *themselves*, the storing and retrieving, are usually unconscious too.

Ordinarily, the process of recollection takes place outside consciousness, "automatically." Memory hands over the names, identities, and recollections we need just as we need them. A friend or colleague walks into the room, a neighbor waves from across the street, and we say "Hi, Nicoletta." We step out front, sniff, and think, "Asphalt—that's right, they're paving Barton Street." When Fred, who works in the next office, says hello in the hall, ordinarily we say, "Hi, Fred" *without* telling ourselves, "I've seen that face before, right here in this area—carrying a coffee mug! *Whose could it be?*" The name is ready for us when we need it. Unconscious memory processes do the job.

Countless times each day we *recognize* pieces of the outer or inner worlds we live in. We know them and know their names. We don't consciously *recollect* them. We simply find them "on the tips of our tongues." To be *conscious* of a thought does not mean we know where it came from. The process of recollection is nearly always unconscious, carried out behind the closed doors of memory. We might picture the conscious mind as a surgeon at work in a hot, narrow spotlight, surrounded by the cool, dark nurses of memory, handing him each recollected fact he needs before he asks for it—before he even knows he needs it. *This* surgeon and *these* nurses work with supernatural intimacy because they are all part of one intelligence, separate districts of one mind.

In grasping that ordinary, everyday recollection happens unconsciously, it helps to notice that when we intervene *consciously* in the process, we usually mess it up.

I was asked for a particular word recently; I rummaged around and failed to find it. The best I could come up with was "compass"—not what I wanted. Ten minutes later I happened to notice the correct word—"calipers"—floating casually in the full light of consciousness, on the rippling-smooth surface of the unconscious, like a lounger on a pool raft, sipping a margarita. As soon as I stopped churning the depths, the right word surfaced by itself. Which is hardly unusual. We've nearly all had the experience of being asked for a name or fact we know well and—under the hot glare of that explicit, specific question—not finding it. Usually when that happens, we have gone *consciously* looking for the recollection we need, instead of trusting the unconscious delivery system that works perfectly most of the time.

Moving Down into the Republic of Being

One important change as we move farther down-spectrum, toward enveloping daydreams and the outskirts of sleep, is implied by "loss of control." It is also suggested by the spectrum's rule of thumb that says, up-spectrum, consciousness feeds memory; down-spectrum, memory feeds consciousness.

Naturally, we become less disciplined and precise in our use of the controls as we move down-spectrum. And the entity we are controlling (or think we are) is increasingly apt to go farther and do more than our instructions tell it to. We leave the memory tap open longer: instead of the relevant items only, entire recollections emerge, including information that's irrelevant to the task at hand. Within memory, each recollection suggests others. As our disciplined use of memory evaporates, we see whole recollections, and then sequences of recollections.

Up-spectrum, the mind is busy acting. Its activities, and the situations they belong to, are copied down by memory and put down in its storage racks. Much information flows from consciousness to memory, more than the modest amount supplied by memory in answering the conscious mind's questions. When it needs memory's services up spectrum, the conscious mind opens the memory tap in a controlled, careful way, withdraws the cup- or teaspoonful of information required, and shuts it down again.

But eventually, moving downward, we can present a query and be met by a barrage of recollections, a sequence that ranges on indefinitely, far past the bounds of an actual response. (By the same token, someone in a meditative, down-spectrum mood might react to a simple question with a teeming flood of recollections.) One recollection leads to another. Instead of just information, or one recollection, or a sequence of five or ten, we get a continuing flow. In Chekhov's story "After the Theater," a young woman sits down to write a letter before going to bed, "and soon her thoughts went wandering, and she found herself thinking of many things: of her mother, of the street, of the pencil, of the piano."

Notice how many of these changes, which deeply alter our experience of mind, are simple, natural results of losing energy, losing acuity, gradually growing tired.

Recall Nabokov's experience: when he is almost asleep, he hears a voice speaking irrelevant words and phrases. That is not so different from the case of Chekhov's teenager, whose thoughts go wandering. Down-spectrum, memory takes over and presents long sequences of thoughts to consciousness. Below or even just *at* the hallucination line, memories are *embodied*: instead of just observing them, we reexperience them.

This is merely a general outline. Different uses of memory are interleaved at every spectrum level. But spectrum level tells us what sort of memory use predominates.

Pause for a Tragedy

The growing importance of unconscious memory processes in determining how we think as we move down-spectrum plays a part in the tragic farce called "transhumanism." Catch transhumanism any day now, coming soon to a university (and then a high school and then a kindergarten) near you. Transhumanism is the idea that, by gradually replacing more and more bits and pieces of the human body with computer chips and other exquisite machines, we will make life better—*and it doesn't matter anyway, because this is the direction of technology and progress*. What kind of pitiful Luddite would ask questions?

Unfortunately, transhumanism is a formula for wiping out the human race. Whether you think *that* is a good idea will determine your view of the project as a whole. Once we have built robots with IQs of 500, we can easily build them with IQs of 5,000 or 50,000; what makes you think that these transhuman wonder machines will keep *you* around, except as a houseplant? One near-term part of the program is to implant memory chips to improve our own organic memories. But no memory chip as we understand such chips today could replace anything but the uppermost, narrowest slice of memory—memory as it operates at the spectrum's top. There, memory is a mere information supplier. But as soon as it starts offering entire, ambience-loaded recollections and sequences of recollections, memory becomes part of your personality. Those sequences reflect both your own experience and the way your memory works: how you move from one recollection to the next, what you notice and don't notice, what you remember and what you omit.

Assuming you had the local surgeon wire the new chips into your brain so that you could choose which memory to invoke—organic or brand-new inorganic—are you sure you would always choose organic when you should? Inorganic is so much faster and more impressive! And with all your competitors at the office running off *their* new chips all the time, could you afford your old-fashioned predilection for having a personality?

The real brilliance and charm of transhumanism will emerge in the way it sets human beings against themselves. Like the Soviet show trials of the twentieth century, in which political prisoners broken by interrogation denounced themselves, not to avoid execution (they were all shot anyway and were never in any doubt about the outcome), but to do their patriotic duty. They believed in the workers' state! Transhumanism, by the same token, will turn each subject into the most persuasive of all voices for his own extermination.

But, as Freud writes at one memorable point in his essays: enough of these horrors.

The Waiting Room

Consciousness, the "conscious mind," has a waiting room that is central to our mental lives. Candidates for consciousness are delivered to the waiting room (what Freud called the preconscious), where we can feel their emotional tone before admitting them to consciousness or turning them away. If they are menacing, we can get rid of them before they enter consciousness. We know they will come back—and we will swat them away again.

In what sense are we aware of such thoughts if we haven't allowed them to become conscious? We can *feel* them before we are fully aware of them, before we inspect them head-on. We feel their contents even if our eyes (so to speak) are averted.

A routine mental event we have all experienced is embodied in the exclamation "*I don't want to think about it!*" "The sun beating down on the deck feels so beneficent that I don't want to admit any thought that might claim me" (that is, that might claim the narrator's attention [Jacob Glatstein, *The Glatstein Chronicles*]). "I don't want to *admit*"—just the right word. A thought *seems* to be right there (in mind, in the "waiting room"), fully formed, patiently waiting for *admission*—to what? Obviously, to full consciousness. Into *attentive* consciousness. I am *already* conscious of this thought: I have it in mind, but it merely lurks on the periphery.

Coetzee tells us the same thing: "An image of Pavel comes back to him"—Pavel, who has just died. "There is something else looming too in the corner of the picture, something he thrusts away" (*The Master of Petersburg*). He is conscious of this something but doesn't want to be.

We need a sufficiently flexible view of consciousness to account for the waiting room. My degree of focus or attention is tremendously important in determining how information reaches me from outside, and what I do with it when it does. By the same token, I can be conscious of something but not attend to it, allow it to sit passively without triggering a mental response (or much of one) aside from *no*.

BLOCKED EMOTIONS

Freud discovered that an emotion conceived but pushed away—an emotion we don't express, never *feel* (a *blocked* emotion, we might call it), frustrated, smoldering, incomplete—is a lit fuse that burns on and on in your mind. It is an unresolved dissonant chord.

Bertha Pappenheim (called Anna O. in the famous case history) was Freud's most important early patient. Once, a friend let a pet dog drink water from her glass. Miss Pappenheim was revolted—but good manners kept her quiet. She suppressed her revulsion. The emotion was *conceived*, so to speak, but never *felt*, never completed, never consummated. It remained in the patient's mind, "undischarged." Frustrated. Blocked.

For Freud, this observation was the start of a journey that took him to a deep understanding of neuroses and repressed memories.[16] Blocked emotions are central, however, not only to psychiatry, but to the spectrum. Among other things, the spectrum is a sort of tug-of-war between thought and emotion. Emotions are increasingly important as we venture down-spectrum, and one of the most important species of emotions is the *blocked* emotion, which goes on expressing itself whenever it can for long periods, or forever.

A blocked emotion is any emotion I have not got out of my system. For example: An emotion I would naturally have expressed had it not been stifled, at the time, by ideas of correct behavior. Or an emotion that I now believe I *should* have felt—love, pity, anger—but at the time, I did not feel. (Maybe because I was a child and selfish, as children are, or because I was afraid to show my real feelings.) Or an emotion I did experience, but not fully: a fire that I hid somewhere while it still smoldered. (Perhaps a childhood scene terrified me, and ever since, I have kept myself from remembering it. But terror needs time to burn itself out.)

Some of these blocked emotions are stored in memory—the fear or disgust, say, that I don't want to remember. Some are merely evoked by memories: the love or pity, perhaps, that I should have felt and showed, but didn't. All blocked emotions, whether suppressed, incomplete, or unborn, are raw and dangerous and psychologically powerful. In a sense, each is unconsummated.

Blocked emotions try to sneak into consciousness. They can be recalled like any other memories—but they find the way barred. Consciousness tries to keep them out. They are most likely to slip into consciousness when your guard is low because you are asleep and dreaming. Even then, a blocked emotion (or the memory that gives rise to one) might have to sneak past in disguise.

An emotion must be carried on through, like a physical action— like a sneeze, a swallow, the swing of a tennis racket, a slap on the back, a leap up three stairs or down four. Most emotions are completed the same moment they are conceived or immediately after. You are pleased the sun came out, happy to smell the coffee, angry at the driver in front for not signaling. All done.

But blocked emotions always find a way to speak. Sometimes they speak in neurotic symptoms—but this is a radical end case. Sometimes we let the dissonance into our consciousness, and it resolves and goes away. But sometimes it cannot resolve. It is bound up with places or people or worlds that are gone.

Our revolted Fräulein Bertha P.—Freud was treating her jointly with his mentor Josef Breuer—had a gift for hysteria. She experienced her blocked dog-disgust as an inability to *drink*, which left her parched and miserable. It was no joke. But when she succeeded in locating the dog memory (she had pushed it out of the way), and she expressed, *experienced*, the blocked feeling of revulsion, that emotion was no longer desperate to make itself known. It receded to normal proportions, and Fräulein P. could drink freely again. The dissonance resolved.

Ordinarily, neuroses don't enter the picture. Blocked emotions are perfectly normal. We thrust a blocked emotion away, as if we were rolling a ball energetically uphill—knowing that it will always come back to haunt the waiting room, waiting to slip into consciousness. A blocked emotion is most likely to make its way past our defenses when our thought is most passive and (potentially) most emotional—at the bottom of the spectrum. There, it often emerges as the center of a theme circle—a series of sleep-onset memories, that is, or a dream.

Strange Thoughts

We can taste and feel the sea before we reach the water's edge. And we can feel the lower spectrum—those strange, vivid, dreamlike states—before we approach drowsiness. How? Why? We don't know. But the phenomena are right there in front of our noses and must be recorded.

The novelist and translator Esther Salaman published an important book on "involuntary memories of childhood"—in effect, brief hallucinations experienced during an ordinary day; waking dreams. They are vividly emotional, joyous or terrifying. She writes, too, about a formidable collection of other authors and memoirists who have described the same phenomenon. Proust's "Salaman memories" are the most celebrated. Nearly his whole literary output is based on those memories.

We don't know how widespread such experiences are, because they are hard to recall; in fact, they are hard to *notice*. But it is clear that they often verge on the experience of pure being.

The recollection is always fleeting, sometimes joyful. You might remember an emotion (*how* you felt) even if you remember nothing else.

You might be overwhelmed by a brief Salaman memory and never notice. "It was usually half gone before my conscious mind seized anything," Mrs. Salaman writes about her first experiences with returning childhood memories. "Often one notices nothing more than a change of mood" (*A Collection of Moments*). Many intriguing mental events happen when we aren't paying attention. But if you monitor your thinking uncritically for a few days, "you will be amazed at what novel and startling thoughts have welled up in you." One Ludwig Börne published this observation in 1823. Freud was fascinated.[17] No wonder! Try it yourself and you might be surprised to find it is true. I did. When these events happen to me, they pass me by completely, leaving no trace of a memory—unless I have primed myself to watch for them incessantly. When you do notice one of these "Börne memories," you will catch yourself in dreamlike thought even though you are fully awake; you will find yourself casually contemplating strange things, bizarre or impossible situations.

By the same token, we are reminded regularly of the past, especially as we move lower on the spectrum. We are consistently reminded, with remarkable precision, of past events that happened surprisingly long ago. And usually we sweep right past. We don't dwell on the unexpected memory. We don't remember having remembered—as if the sidewalk underfoot suddenly revealed a chink through which you could see miles into the molten depths. But when we don't remember, it never happened.

Then we might look at, for example, the New Zealand novelist Katherine Mansfield's celebrated "Bliss," which is full of brief, blazingly vivid emotional experiences that take place during an ordinary late afternoon and evening. A dream equals hallucination plus emotion. The brief experiences that "Bliss" describes are enormous swells of emotion, as in dreams, but without hallucination. They are not "waking dreams" in the simple sense. But you would never mistake these moments for normal. They are transporting.

Adding up all the evidence reveals that mental life—for some of us, maybe all—glitters with dazzling foam specks thrown up by shattered dream waves. Flecks of dream touch us like faint spray far from the ocean's edge, as the light diamonds thrown by stained glass do when the sun shafts align just right. We rarely remember them. Often we don't even notice them. (How could we *not notice* such vivid experiences? In overconsciousness, again, we are so caught up in sensation and emotion that we have no attention for anything else.)

"Involuntary memories" are a broader group that include Salaman memories, but involuntary memories don't necessarily have dreamlike characteristics. They are merely recollections that come to mind unsought. They are routine, the psychologist Dorthe Berntsen believes; and "some two thirds of all involuntary memories . . . occurred when her subjects reported being in a nonfocused ('diffuse') state of attention."[18] Clearly, they are a low-spectrum phenomenon. But we don't know whether Salaman or Börne thoughts happen to everyone, or only to low-spectrum personalities. We don't know whether they happen at any time, or only at the bottom of down-spectrum oscillations. Mrs. Salaman's own experiences happened, usually, when she was "sitting in my study, blind and deaf to sensations." Which tells us little. Proust's are the most famous of all such memories. The originals, on which he based his monumental novel, did, in fact, strike him on a winter's evening as he came home freezing out of the snow ("Contre Sainte-Beuve" ["Against Saint-Beuve"]). Shakespeare's *Winter's Tale* (and a million other witnesses) remind us that winter, as well as evening, makes us look inside, to the inner field of consciousness and the storytelling end of the spectrum.

In any case, "while awake," the neurophysiologist J. Allan Hobson tells us, "dreaming is essentially impossible." Wrong—as Wordsworth knew long ago.

The Great Spectrum Struggle: Lust versus Danger

We are drawn to all remembered scenes of vivid emotion—especially to memories of pain or violence. Straight vodka to an alcoholic, heroin to a junkie, equals strong emotion to an ordinary person. We *want* to remember and think about scenes that evoke strong emotion. "They were drawn by the dark curiosity that trawls minds to the grotesque" (Cynthia Ozick, *Heir to the Glimmering World*).

They draw us because of our morbid fascination, and innate sadism or masochism. (We are sadists insofar as we are drawn to other people's pain, and masochists insofar as we are drawn to such memories even though we find them unpleasant or painful or revolting.) Sadism and masochism are common as dirt, human as breathing. But we are drawn to these recollections most of all by their sheer *intensity*, their brightness in the night sky of memory. The brightest, loudest, biggest, boldest *anything* invites our attention. Naturally, our attention is drawn to these strong memories steeped in emotion.

We are attracted to nearly *anything* that will make us feel strong emotion, especially in the sensational world of the twenty-first century. We are drawn (obviously) to pleasure, and we construct daydreams and fantasies to feel pleased. We are strangely drawn to unpleasant emotion too. Faced with violent or horrific scenes, we don't want to look. But we *must*. And then we draw back in horror.

We might feel like Desdemona after hearing the terrifying war stories of her secret fiancé (soon husband), the great warrior Othello: "She swore, in faith, 'twas strange, 'twas passing strange, / 'Twas pitiful, 'twas wondrous pitiful: / She wish'd she had not heard it." But she always asked for more.

Return to the Waiting Room

I have already referred to the "waiting room," a critical function of mind. Facts are not foisted on us blindly; recollections and ideas are delivered from memory into a "waiting room" just outside consciousness.

The conscious mind knows what's in there. We can feel it. And we are free to accept or reject these "thought candidates." If they are painful or upsetting or merely inconvenient, they are booted out of the waiting room and go back to memory. Memory proposes; conscious mind disposes. But in the end, we cannot forever evade an important memory, a memory loaded with a substantial emotional charge. If we refuse delivery, it will slip into dreams or down-spectrum daydreams. If we refuse even to *dream* about it, it will probably reemerge as a neurotic symptom.

Obviously, we feel differently about pain and suffering if it is *our* pain and suffering. But not terribly *much* different. In fact, very little different! Again, we are ambivalent. The memories are painful, but we are drawn to them. The phrase "don't dwell on it!" exists because we *do* dwell on embarrassing, unpleasant, painful memories. We dwell on them out of self-fascination, sometimes prudence (figuring out how to avoid these things in the future), or a desire for vengeance, but mainly because we get a kick out of them. Of all vivid memories, these are the most vivid.

These personal-suffering memories make for intense internal struggle—to remember or not?—as we move down-spectrum. There are three categories of trouble: (1) bad, (2) dangerous, and (3) banned outright. Bad memories can slip into ordinary, waking consciousness, especially down-spectrum. Dangerous memories can usually enter consciousness only when we dream. Banned memories are the ones we repress. We try to get rid of them entirely.

Dangerous memories often involve childhood pain and suffering. They are the most intense and therefore most frightening. The restrictions imposed by dreaming—ideas and recollections must be presented *visually*, are made to fit an ongoing theme-circle narrative, *and* could be distorted by the organic process of recollection when we are asleep—sometimes cause these memories to appear in disguise. But we nearly always recognize them anyway.

Yet some of these memories (the absolute worst) are so upsetting that we will never remember them. We repress them. They wind up stuck in the mud floor of memory, never to rise again unless we make a point of digging them out. ("These flashes of illumination disturb him; rather than holding on to them, he tries to bury them in darkness, forget about them" [Coetzee, *Youth*]. "When we start digging around in our souls," says Anna Karenina's husband, "we often unearth something that might have lain undetected" [Tolstoy, *Anna Karenina*, Bartlett translation].)

But these dark memories aren't gone. Freud tracked them down, as I have mentioned, in the form of neurotic symptoms.

BRUTAL MEMORY

Macbeth consults a doctor about his wife. She had urged him to murder the king and seize the crown, and she helped him do it. But now, that merciless murder is unraveling her mind and dragging her downhill toward suicide. Macbeth knows that her obsession with these guilty memories is driving her mad. He loves his wife. He asks her doctor:

> Canst thou not minister to a mind diseased,
> Pluck from the memory a rooted sorrow, . . .
> Which weighs upon the heart?
> (Shakespeare, *Macbeth*)

Answer: *no*. There is no way to "pluck from the memory a rooted sorrow." The doctor's answer: "Therein the patient / Must minister to himself." Basically, that was Freud's thought too.

Our penchant for recollections that trigger strong emotion is just one expression of our consuming lust for emotion. The most *direct* expression is our fondness for the thing itself. Many people love horror films. Many don't; but anyone who goes to the movies is probably seeking strong emotion. Love of roller coasters might not be widespread, but we nearly all engage in activities designed to flood the senses, from music to sports to sex. Such activities rarely approach overconsciousness or pure being (except for sex, in certain circumstances). But they all point in that direction.

Sadism and masochism underpin some of our greatest artistic

achievements. *King Lear* includes one of the most horrifically violent scenes in all of drama: the on-stage blinding of Gloucester. Hugo's *Notre-Dame de Paris* [*Hunchback of Notre Dame*] and Dumas's *Comte de Monte-Cristo* [*Count of Monte Cristo*] are probably the most popular of all French novels. The first makes effective narrative use of the torture of a beautiful young girl, plus many executions and murderous fighting. The second includes an unforgettable (unfortunately) account of the most disgusting (as far as I know) public execution in all literature. Both are not only popular but great novels.

To reason is human. To long for our minds to be flooded with powerful emotion, so that we can only feel and can't think, so that we *can't* reason, is also human. We long for pure experience. We long to lay down the burden of reason as kings once longed (they said) to lay down their crowns for an occasional rest period. Reasoning is the crown jewel of human achievement, but it is hard work.

Approaching the Hallucination Line

We know that to "sink into yourself" is one way to understand the movement down-spectrum. Consciousness shifts its main focus from the outer to the inner fields. We need to rest, to leave our bodies in peace, and (not insignificant!) to close our eyes. So, naturally, we "withdraw consciousness" from the body and its perceptions of the outer world. The inner field must occupy us henceforth. And naturally, the inner world glows brighter as the outer world darkens and disappears.

But if we don't mistake our memories and ideas for reality when we are wide-awake, why make that mistake when we are asleep? Why a hallucination line? Because (among other things) something tells us that external reality is always there, even when our eyes are closed and we sleep. Since it *is* always there, we expect it to be there whenever we are conscious. When we are conscious but *asleep*, and the outer field has blinked off, it's natural to reinterpret the inner field as reality. After all, we don't believe that reality has disappeared merely because we have fallen asleep. So *where is it?* Hallucination steps into the breach, and we stay true to the continuity-of-reality assumption.

Mental Dusk and Sleep-Onset Thought

We have been following the spectrum downward, into mind wandering and daydreaming—into the world of emotion and sensation. Let's continue.

As mental energy continues to fall and focus continues to grow more diffuse, sleepiness begins. Daydreams grow more distracting. Or we stare out a window, and thinking just stops for a while. ("He sometimes locked his fingers behind his head, closed his eyes, and emptied his mind, wanting nothing, looking forward to nothing" [Coetzee, *Life & Times of Michael K*].) Eventually, we find ourselves free-associating as we slide toward sleep and dreams. The fact is not as well known as one might guess, so, for the record: you *will* find yourself free-associating as you approach sleep.

There is a good reason why the French *songer* should mean both "think" and "dream." *Songe bien, oui, songe en combattant, qu'un oeil noir te regarde*, Bizet's toreador sings in the famous lyric from *Carmen*. "Think well—*dream* well—while you fight; a pair of black eyes watches you. And love awaits." Thinking and dreaming can blend into each other. But there is another reason to cite this wonderful song. A moment's thought will convince anyone that nobody becomes a matador to get girls. Less dangerous tactics are available. Many people find it impossible to grasp that some men *want* to fight bulls for the fight's own sake—and only then for the danger, the triumph, the fruits of victory. The *Carmen* lyric, by Henri Meilhac and Ludovic Halévy, tells us plainly that soldiers and toreros are the same breed, insofar as "for pleasure, they do battle." But that's only the verse. By the chorus it is forgotten, and the watching-eyes theory steps forward.

The boundary between thinking and dreaming is a fascinating, little-known part of the mind. We all know that dreams are hallucinations, but the special character of the "sleep-onset mentation," or hypnagogic thought that leads to sleep, is not widely known. We know that it is loosely associative or free-associative or maybe a "stream of consciousness." (William James introduced the term "stream of consciousness"; Wordsworth had already written about "the river of my mind" [*Prelude*, Book 2].) But it's not just that; it is often a series of separate, short-term hallucinations. "Pensive awhile she dreams awake," says Keats of his perfect specimen Madeline, who is half-undressed and ready for bed ("The Eve of St. Agnes"). To "dream awake," to experience dreamlike hallucinations while still awake but on the verge of sleep, is the exact character of sleep-onset thinking.

I once embarked on sleep onset with the following unimportant but typical experience. I thought I was fully awake; I believed I was merely *thinking* about holding a coffee mug. But suddenly the mug seemed to slip out of my hand and fall—which startled me alert. I realized that I was hallucinating and had been nearly asleep.

If you are interrupted shortly before dreaming or sleep begins, your own account of your just-interrupted sleep-onset thought might surprise you. "I was at Mr. Schwartz's front door and he was saying, 'Why not come visit on Tuesdays as you used to?'" You weren't thinking about or recalling that encounter. You were *experiencing* it. You stood at the man's door. You heard his voice. (Perhaps he died years ago.)

Still, you were able to regain awareness and awakeness more easily than if you had been properly asleep or dreaming.

Had you not been interrupted, your next thought or hallucination would likely have been (or *seemed*) unrelated. Usually there is no obvious story line to sleep-onset thought. But each step along the way can be wholly enveloping.

Below is a typical (so it seems) sleep-onset sequence. In discussing sleep-onset thought, I am especially dependent on my own logs. It's hard to find substantial examples either in literature or in science.[19] I want, also, to show continuity between sleep-onset thought and dreaming (I will quote dreams in a later chapter). To do that requires dream and sleep-onset examples from the same source.

Sleep-onset sequences are a mixture of ordinary thoughts and hallucinations. The proportion of hallucination to ordinary thought seems to increase as we approach sleep. In this example, each element was a hallucination.

[1] Our macaw stretching his left wing. [2] Ping-Pong. [3] The iridescent colors on a pigeon's gray neck. [4] Rabbi S. in a car; he is driving. [5] Rain on Yale campus. [6] The smell of spring rain on campus and on the streets of Flatbush, Brooklyn. [7] C.'s long, dark, silky fragrant hair; Palestrina, Rilke. [8] A feeling of turbulence in which many memories are dissolved.

Here's an abridged guide to the complex associations in (merely) this one short sequence: I had recently seen the macaw stretch his wing. Elements 1 and 3 are related: colorful birds. Elements 5 and 6 deal with urban rain. The underlying theme of this sequence is clearest in element 4: as a student, I had invited Rabbi S. (a brilliant young scholar) to come to Yale for a debate on biblical issues. (In fact I was driving; I think it *was* raining, but I'm not sure.) The theme is conventional: the comfort and safety of home versus the risk and excitement of college. Here, "home" is my grandparents' home in Brooklyn. Rabbi S. and I were driving between "home" and college.

At the time that this sequence came to me, I hadn't experienced the fragrance of spring rain in Flatbush for thirty years. But the past reaches out to us. Inevitably we accept its advances, lose ourselves in its embraces—remember nothing.

Furthermore, my college roommate and I would often play Ping-Pong in late evening after we'd done some work. Across the quad in our dorm complex was a small room with a table. The visual, sensory aspect of every memory seems to be emphasized; "Ping-Pong" was (to my mind) colorful: a forest-green table, bright vermilion padding on the rackets, walls of the small room painted some vivid color—amber, I think. Element 7 took place at college on my birthday, and Palestrina and Rilke refer to birthday presents from my roommate in earlier years; C. has something to do with them also, but I'm not sure what. The last, unclear element (8) had a feeling of mist that brings to mind a sentence from Dumas's *Comte de Monte-Cristo* [*Count of Monte Cristo*]: "His mind floated like a vapor, unable to condense around a thought."

There is far more to this story, but the bare outline is good enough for now. Perhaps it's clear that these memories form a theme circle.

One other, very simple example is this single hallucinated scene:

I am swimming in the YMCA pool in Huntington on Friday afternoon, thinking about the school Christmas play (I played violin in the orchestra) and the chorus singing a particular song (hearing the tune and lyrics), looking fondly at V. among the girls in the chorus.

I would have been twelve when the original scene happened. Again, I wasn't *remembering* that swimming pool; I was in it—and knew exactly what time of week it was and where the calendar stood. I was crazy about V. This happened during a year I rarely think about; my family had moved and I had switched schools and lost friends. A hard year. Unhappy. This sleep-onset thought alerts me to perverse feelings I must have developed long ago and rejected for their unorthodoxy, for not fitting my agreed line. In fact, that long-ago year wasn't *all* bad. I wouldn't have believed that I had a memory of anything like this scene, in anything like this degree of detail, until I recorded this sleep-onset event.

Dreaming Is Remembering

Rappelling down the cliff face of consciousness leads us, at last, to dreaming and a simple truth. The truth is neither new nor difficult, but it is important.

When we dream, the inner field of consciousness (imagination and memory) dominates the outer field. Memory feeds consciousness, and we are at the bottom of our ability to *control* consciousness, to decide which thoughts enter and which are turned away. All this means, in sum, that *dreaming is remembering, unconstrained*. Ideas and speculations appear too, expressed in visual form, but remembering dominates dreams. Freud knew it: dreaming, he wrote, is simply "another kind of remembering."[20]

"The moment I had fallen asleep I was woken up again by a great feeling of terror" (Karen Blixen, *Out of Africa*). She has *remembered* something she had refused to think about during the day: a defenseless young gazelle, bound, being offered for sale by children at the roadside. She had seen it but refused to think about it. She had set the emotion briskly aside, added a blocked emotion to memory.

Asleep, with her guard lowered, she admits it to consciousness as she dreams. She is *remembering the emotion she had ignored.*

Anna Karenina was unfaithful to her husband. Whenever she thought about it, "she was overcome with horror and drove those thoughts away"—just as Karen Blixen drove away her thoughts about the helpless gazelle. "But in her dreams, when she had no control over her thoughts, her position presented itself to her in all its hideous nakedness" (Tolstoy, *Anna Karenina*, Bartlett translation). Dreaming is remembering.

Spectrum Law: Dreaming is remembering, out of control.

We start with recent memories and work our way back. In the process, we discover what truly interests or worries us. We are good at rejecting unpleasant thoughts, keeping them out of *waking* consciousness. Even in dreams we never surrender completely; dreams tend to be haunted by "dysphoria," unfocused unhappiness. The waiting room holds unpleasant memories that we can *feel* but will not allow into consciousness, even when we dream. Asleep, however, we are not careful enough to be consistent. We let dangerous thoughts slip by. We have nightmares.

Now the remorseless engine of memory is in charge. The doors of memory are wide open and we feel the night breeze playing. We can no longer protect ourselves from dangerous or frightening thoughts. A bad memory can be on us like a famished bear out of the dark before we can turn and fight. "A heavy summons lies like lead upon me, / And yet I would not sleep," says Banquo, Macbeth's close friend and fellow warrior. He is exhausted. He is courageous. But he is afraid of sleep. He anticipates bad dreams; he can *feel* them coming. "Merciful powers," he continues, "Restrain in me the cursed thoughts that nature / Gives way to in repose!" (*Macbeth*).

Other people can guard your body, but no one can guard your

mind. By allowing yourself to fall asleep, you are standing down, relaxing your own guard *almost* completely and hoping nothing bad will happen. But sometimes you know it will.

In dreaming, the conscious mind has left the memory tap on, and recollections flow. One thing leads to another. Frightening or painful memories approach. Up-spectrum, the conscious mind would feel a dangerous memory thread and shut it down. But in dreaming, the gates are open. All the conscious mind can do is to improvise a continuous narrative (or try to) using the material it is given.

Hobson summarizes dream thought as "illogical, bizarre."[21] A widely held idea, which (I believe) is not quite true. Our conscious thought when we dream, as when we are awake, is a *rational attempt to make sense of reality*. But what *is* reality? When we sleep, the *inner* field of consciousness is reality—and presents us with a series of recollections that probably make no sense as a sequence *and* might each be damaged or distorted. (Damaged or distorted because of the state of our sleeping brains, or the tendency of memory—out of control—to present several recollections superimposed.) In short, it's not that our *thoughts* are irrational and bizarre when we dream. *Reality* is irrational and bizarre! Making sense of *this* reality is a stiff assignment, but we do our best. We do it by inventing theme-circle narratives—because at the spectrum's bottom, that is our technique for making sense of the world.

In another sense, too, dreams are *not* illogical. As strange as the narrative might be, the underlying emotions are usually clear when we are awake and thinking about the dream. The *theme* of a dream is an emotion, or an emotion-steeped image. Sometimes it's an unpleasant or painful emotion that we refuse to think about when we are awake. But when we dream we are careless, and our reach into memory goes far. That combination brings to dream consciousness emotions that bother us, that we ordinarily refuse to think about.

In looking at the dream itself, we note a point of agreement between

Freud and the anti-Freud, J. A. Hobson. (There are many anti-Freuds, but Hobson will do for now.) Dreams are concrete and *visual*. In dreams we have no ability to invent language, although we can recall and understand it. We *can* invent pictures. Dreams speak in images.

Dreams, however, can't simply present disembodied emotions. A painful emotion gathers pictures to make its point, and the pictures might be illogical or confusing or absurd. But we can usually detect the underlying emotion when we are awake and able to reflect—when we have our *selves* back. And those emotions tell us the truth about our minds.

The psychoanalyst Stephen Grosz spends his time analyzing other people's dreams, but he also reports one of his own.[22] In this dream he reaches for a small lizard disappearing between two rocks into the earth, but he cannot catch it. It's gone. When he wakes, four letters linger in his mind along with the image: *S, I, D, A*. He works out the dream's meaning by free-associating from the remembered fragments. Years earlier, as a young clinician, he had met with a patient who refused to be treated for a disease associated with AIDS. He had tried to convince the man to accept treatment and *had* convinced him to take better care of himself—but only by heading off early for a holiday in Rio, not by checking into a hospital. Several months later the patient died. This dream, decades later, was an image of that decades-ago, deeply upsetting event and the dreamer's reaction.

The AIDS patient had grown up on a peninsula at the tip of Cornwall, called the Lizard—the southernmost point in England. The Spanish Armada had first been sighted, the patient had remarked, from a field next to his childhood home. *SIDA* is Spanish for "AIDS." The young psychoanalyst had tried his best to save the man—who had slipped through his fingers, who was dead and in his grave. There is more to this simple dream. But its eloquence (meaning Grosz's eloquence, of course) should be clear.

Grosz's unconscious wants to remind him: decades ago you lost that patient, whom you might have saved. (These are not facts; they are just unconscious mutterings.) But dreams can't speak English. They must speak pictures. The barely escaped lizard darting underground, into the grave, is the result.

Old stories are like dreams. They speak concretely, *visually*, in images. In Shakespeare's *Julius Caesar*, a wild thunderstorm is understood by Cassius as the image of Caesar himself, who is dangerous and must be got rid of. I could, says Cassius to Casca, "name to thee a man / Most like this dreadful night, / That thunders, lightens, opens graves, and roars."

This passage is an introduction to a still more remarkable one in the Hebrew Bible—one of the best-known in Western literature, and least well understood. Because we read it at the wrong spectrum setting, we do not understand it. Intellectual and literary habits and fashions have changed decisively, leaving us to understand the words but not the sense. (I have discussed the story in detail elsewhere.)[23]

Moses has escaped from Egypt to the calm of provincial Midian, where he is tending his father-in-law's flocks in the wilderness. He is brought up short by one of literature's most famous visions: a burning bush that is not consumed. It simply burns on and on. A divine voice from the flames tells Moses he must confront the most powerful man in the world, the emperor of Egypt, and demand freedom for his enslaved fellow Jews to make their way back to their ancestral homeland.

If the majestic vision is intended merely to stop Moses in his tracks and shake him so deeply that anything is possible, it does its job. But that's only part of its purpose. Its most basic task is to reveal Moses to himself, show him the man he really is, convince him that there is no hiding from his own sense of justice.

The burning bush is a picture of Moses's character, a visual synopsis of his psyche. Moses *himself* is the thornbush aflame with passion that never burns itself out. (Passions burn in Hebrew as well as English.) We know Moses is a passionate man: he was forced to escape Egypt because he had come upon an Egyptian beating a Hebrew, and in his towering rage he had killed the Egyptian with one stiff blow of his heavy stick. In Egypt he is now wanted for murder. At the same time, a thornbush is the lowliest of trees, worthless desert scrub (ordinarily, it would burn to ash in moments)—and Moses is called the humblest man on earth (Numbers 12:3). He describes himself as incapable of confronting the pharaoh and leading Israel. But his unwavering passion makes him powerful and dangerous, like the burning bush.

The vision is a prediction also, not just a statement of fact. It has no proper end—the bush burns on—and Moses himself dies without reaching his goal, overlooking the land of Israel without ever having set foot there. (This is one of the Bible's most remarkable comments on the nature of human passion.) Moses dies with his passion unconsummated. Like the bush itself, he never calms down, cools off, comes to rest.

Thinking in images is more natural to biblical civilization than to us; words in the Bible are most important as bearers of images. (The images of the Hebrew Bible—from the dove with the olive branch in its beak to the lion lying down with the lamb, the chariot of fire, the writing on the wall, and many others—are the basic images of Western art and literature.) Images occur in the Bible not to underline or decorate a point; they *are* the point. Words are only the medium in which they are presented. Thus, too, for example, we fail to grasp that an all-night fight with an angel and a vivid dream are virtually the same thing. The dividing line between dream and reality is far more lightly drawn in the Bible than it is for us. The Bible's world is a down-spectrum world. The visual, subjective reality of the lower spectrum is the Bible's native landscape—and the theme circle is its basic narrative form.

One of the most striking things about dreaming is that we don't have more nightmares than we do. Anyone who's ever had even one has *some* terrifying memories. Those bad thoughts are able to push their way into consciousness when we are dreaming, when focus and control are at their lowest. And our inability to control memory when we are dreaming increases the likelihood that we will pull up a monster memory from the deep. "Don't dwell on it"; we are drawn to violent memories, even as they repel us.

Why Dreams Predict the Future

Shakespeare's Duke of Clarence is the future king's brother in *Richard III*. The future king is a murderer. Naturally, his brother isn't eager to acknowledge that fact. He tells everybody, including himself, that Richard loves him. Clarence *knows* that Richard would murder him without a second thought, but refuses to admit it—least of all to himself. He conceives the feeling that Richard will murder him as a matter of course—but he rejects it, refuses to *feel* it, and adds these blocked emotions to his unconscious mind.

As he descends the spectrum, Clarence's mind (like all minds) opens up, drops its guard. Clarence dreams what he has on his mind but has banned from his waking thought. He *feels*, in his dream, the emotions (terror, hopelessness, shame) that his conscious self refused to feel. He feels that Richard will kill him.

And, of course, Richard does.

Why do dreams predict the future? Because they tell us truths we know but are not brave enough to acknowledge. They don't so much foretell the future as remind us what it was always going to be. "I feared those dreams. They swarmed like reenactments of something foretold" (Cynthia Ozick, *Heir to the Glimmering World*).

The Only Protection

The mind does protect us from the frightening content of dreams, by the only means it has: making us (or letting us) forget. There is nothing else it can do. On the other hand, the mind *can* reverse time, unwrite unreality, and, under narrow but important circumstances, foretell the future.

The mind, in sum, follows a great tidal motion. At its logical peak, reality and self are two separate things. Our reflective selves and the reality on which they reflect are different. But from the start of our journey down-spectrum, the borders begin to blur. And at the end of the trip, our real selves have been absorbed into dream reality, and only our hollow unreflecting dream selves are left on the narrow ledge of consciousness—the place that remains after dreaming has taken what it needs. Reality and self have both changed radically from what they were.

And in the morning? Reality, rubbing its eyes, moves out into the external world again, leaving the reborn, full-fledged *self* cool and quiet, slightly dazed—ready to start over.

Four

A Map

I have outlined the spectrum in several ways from overhead, but many points require a closer look—from the ground. We need to follow examples of up-spectrum thought, of the mind acting normal in midspectrum, of daydreams and fantasy and sleep-onset sequences and actual dreams. I have made heavy use of the terms "conscious mind" and "unconscious mind," "memory," "thinking," "feeling," and others. We don't need to analyze each of these topics. By and large, the mainstream research community is working that project—from many directions, with staggering resources wielded by some of the finest thinkers in this particular universe. But we need a concise overview so that we can use the basic terms with confidence. In some ways, the map I require is simpler than any mainstream map.

I will concentrate on *consciousness* and *memory* and the two basic facts of conscious mind: acting and being—*thinking* and *feeling*. My basic map of mind has two regions: (1) *conscious mind* and (2) *unconscious mind*, otherwise known as memory. Conscious mind divides into *thinking* and *feeling*. There are additional subdivisions, but not many.

Mind is consciousness and memory. Consciousness deals only with *now*; memory, with *not-now*, with the past. I can think consciously about the past or future, but I can *experience* only the present moment and no other. When I reexperience the past in sleep onset or dreams, the mind brings the past *to* me, brings the past to the present rather than sending me outbound to find it.

Consciousness anchors me to time—fastens me to the relentless forward drag like a cable car gripped onto the moving steel rope beneath the street.

I *am* conscious of a sort of fading afterimage of the past moment or two, and the near future in my headlights, just about to happen. Edmund Husserl points out that I am always conscious of where I have just been and *where I am going*. This knowledge of where I am going is no matter of deliberate plan making. It is the mind's *glancing forward* along the trajectory of behavior that is already established. And as such, it is important.

How do we invent dreams? For some of us, storytelling is natural and easy; for others, it's nearly impossible. Yet *we all dream*. How do the non-storytellers manage it? The fact is that, when we improvise dreams, clearly it is the *near future* in the Husserl headlights, our knowledge of what must be coming up next, given our behavior and experience, that guides the dream-plotting function. In waking life, we can foresee what's (almost) bound to happen in the next half minute (or even the next few minutes, depending on circumstances). This knowledge, our mere *mental momentum*, is what guides the production of our dreams.

Memory, for its part, deals only with not-now. I can't recall something *while* it happens—cannot create and recall a memory simultaneously.

In this sense, one might say that consciousness and memory are orthogonal, like the horizontal and vertical beams in the frame of a skyscraper. They exist at conceptual right angles. Consciousness deals with objects and events that cannot *now* be part of memory, because they are only just happening. Memory deals with items I cannot *now* just be growing conscious of, because they have already entered memory.

Mind-Map Principle: Consciousness and memory are orthogonal, in the sense that consciousness can deal only with the present, and memory, only with the past. I can *experience* only the present moment. I can *remember* only the past.

Whatever is in memory, I have already been conscious of, at least once before. In this sense, there's nothing surprising in identifying memory with the unconscious: one expects the conscious and unconscious minds to exist at the same level, as orthogonal alternatives.

Years ago I was giving a talk on these ideas and needed a concise picture of the mind's changing relationship to the outer versus inner worlds. I was attracted to a picture that was simple and useful, and even though it appeared to be fundamentally flawed, I had to use it.

The picture showed a sort of minimal, two-zone archery target: the "up-spectrum" image had a bull's-eye labeled "Memory," surrounded by a ring labeled "Conscious Mind"—surrounded by the whole outside world. Up-spectrum, conscious mind can turn to the outer world on one side and memory on the other. It is sandwiched neatly in between.

The "down-spectrum" image had the regions reversed: conscious mind was the bull's eye, surrounded by memory—surrounded, again, by the world at large. When we sleep, signals reach the conscious mind *only* from memory. Signals from the outside world reach sleeping consciousness from a distance (so to speak)—from the other side of memory, from the "far side"—only when they are strong enough to travel right *through* memory. Proust describes a valet speaking to the sleeping Swann: as the valet's words "plunged through the waves of sleep in which Swann was submerged, they did not reach his consciousness without undergoing the refraction which turns a ray of light in the depths

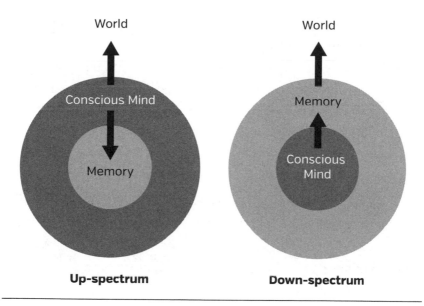

FIG. B.

of water into another sun" (*À la recherche du temps perdu* [*In Search of Lost Time*]). External reality needs to penetrate deep to reach us when we sleep.

During the trip down-spectrum, consciousness and memory trade places. The outer ring becomes the bull's-eye, and the former bull's-eye is transformed into the outer ring. This is a simple way to describe the basic transformation in the mind's structure brought about by the spectrum.

This seemed like a good and simple model. The problem was that conscious mind and memory were a false pair—socks and kumquats. They did not seem like *counterparts*, like two elements on the same conceptual level. It seemed arbitrary and strange to say that, if you stripped mind down to the basics, you found conscious mind and memory. Yet otherwise, the picture seemed right.

Only years later did it occur to me that memory *is* unconscious mind. Now the picture makes sense. Conscious mind and unconscious mind are obvious counterparts. We merely need to remember that unconscious mind is memory.

Having plotted the two basic regions of the mind map, we can go further—if only a little. Conscious mind contains mental states that occur on a spectrum from thinking to sensing and feeling. These two are also (in another sense) orthogonal: you can think one thing and feel another independently, at the same time. But you cannot think about two different topics simultaneously. You *can* experience many feelings or sensations at once, but they blend together like drops of colored ink in water. In this sense, many separate feelings become one.

Continuing in this way, we can divide thoughts into perceptions, recollections, and ideas.

Many other schemes are possible too. Consciousness can be understood in terms of outer-field versus inner-field events. We could also think of consciousness as a point traveling through time with memory trailing behind it, reaching back into the past—as if consciousness were a comet and memory its tail.

An old tradition divides mental or psychic reality into reason and will, cognition and volition. Reason deals objectively with reality,

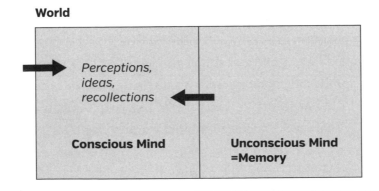

FIG. C. *Basic mind map.*

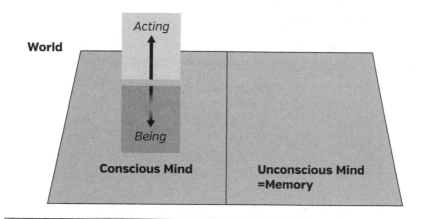

FIG. D. *Basic mind map (1).*

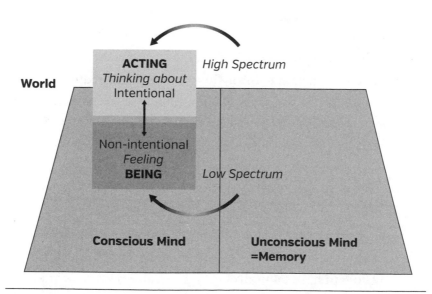

FIG. E. *Basic mind map (2).*

or at least aspires to. Will reflects the subjective reality of a thinking creature. But for our simple, practical purposes, this is a wrong dichotomy. What's important to a practical view of the mind is that it can think and can sense and feel, can *act* or just *be*. Volition often involves elements of thought *and* feeling. I will raise the idea again, but it's not central for our purposes.

Thinking

"Thinking" is rarely used as a technical or philosophical term. A thought is a mere "pre-theoretical umbrella term," says the philosopher Jesse Prinz.[1] But thinking has a clear intuitive meaning: the conscious, deliberate manipulation of mental states to achieve a particular goal, given certain raw materials. The flagship case is reasoning. I start with certain givens (I am on a Manhattan sidewalk, am thirsty, have money, see a hummus-and-soda cart on the corner) and a goal (to satisfy thirst). Then I lay down a logical pathway that gets me from givens to goal—a *mental* pathway, a sequence of thoughts. Converting this mental pathway into action might require another round of reasoning. This is mental manipulation, mental *doing*, the bringing to bear of mind on reality.

Just what are those "conscious mental states" that we assemble in the process of thinking? We might call them, yes, *thoughts*! (Of course, I already have called them "thoughts.") For our purposes (these definitions are brief and informal), a thought is a *perception* ("the rain has stopped"), a *recollection* ("yesterday was dismal"), or an original concoction, a product of the imagination—I'll call this an *idea* ("Plato will have to forgive me"). I can also perceive something that overwhelms me; a lion slips out of an alleyway and comes roaring down the sidewalk. I stop *thinking about* and am placed in the dangerous state of pure being, or at least in a state where feeling dominates thought, until I pull myself together. (Quickly, we hope.) Or I can recollect an overwhelming memory, or invent an overwhelming idea. Nonetheless, *thoughts* are anchored in perceptions, recollections, or ideas—even though feelings or even pure experience can also grow from these sources.

The distinction between recollections and memories is important. A *recollection* is an event. A *memory* is a snippet of experience, or a fact or rule or something else we have learned, available for recollection. Any thought—perception, recollection, idea—is a mental *event* in time. We can remember having a thought just as we can remember having a sandwich.

A thought, as I have already noted, must be *about* something; it has intentionality. I have "more offences at my beck than I have thoughts to put them in," says Hamlet, breaking it off with his girlfriend with all his customary tact: by telling her how tremendously lucky she is to be rid of him and, by the way, that she should put herself in a nunnery right away, lest she have children and pollute the world. He is indeed a perfect jerk. But she is in love with him. Never mind; the remarkable thing about this casual statement is that it describes thoughts as *containers* for putting things in. That is exactly what intentionality means: if you believe John Adams was a great man, then your belief *refers* to something, is a basket that (abstractly) contains John Adams.

Sensing and Emotions

Sensations and emotions are not mental containers. They are ways to be. *Sensations* and *emotions* are different but closely related. "My foot is painful," "that thought is painful": the first is called a "quale," or qualitative sensation; the second, an emotion. These two pains are different, but they are both literally pain. They create the same cringing urge to escape, can dominate all other mental states, and cause the same exhaustion and despair over the long term. Sensations (or qualia) and emotions are instances of *one* class of mental phenomena. Just as there are several types of thought, there are several types of feeling.

We can see the closeness of sensation and emotion in a few simple examples.

Consider a recollection: "Today is Valentine's Day." Your emotional state might be happy, sad, embarrassed; in plain English, you *feel* happy, sad, or embarrassed. The word "feeling" hints at the continuity between "I *feel* the fuzzy flocking on the Valentine's card" and "I *feel* embarrassed that I forgot to get you a present." Notice the qualitative closeness between a *sensation* such as "this ice-patch feels slippery underfoot" and the *emotion* of unease or anxiety. The two feelings are obviously related, are similar—and not metaphorically! The *actual feelings* are similar, in the sense that bricks and concrete feel similar to the touch.

"Feeling" and "emotion" are synonyms, in the right context, because emotions are grounded in the body. "Emotion dissociated from all bodily feeling is inconceivable," writes the eminent late-nineteenth, early-twentieth century philosopher and psychologist William James, older brother of the great novelist Henry. "A disembodied emotion is a sheer nonentity."[2] He was a radical on the subject, but one doesn't have to follow James very far to notice (as I mentioned earlier) that pure emotions, like happy and sad, have obvious physical expressions. When you are happy, your body feels different from when you are sad.

There is no clear dividing line between mental and physical where feelings are concerned. In discussing Freud's invention of the "drive" (as in sex drive), Jonathan Lear writes that the concept "may even call into question the idea of a sharp boundary" between mind and body.[3] The deep mutual interpenetration of mental and physical, interlaced fingers, is a foundational fact of psychology in general.

Thoughts are always about something, are intentional states. Feelings are ways of being and are *about* nothing; they are *not* intentional states. Language helps us see this subtle distinction, which is easy to miss.

We can say, "I believe that we're leaving" and "I'm sad that we're leaving." It seems as if we have just heard about two intentional states—one belief and one emotion, each *about* or *referring to* the same thing. But language, having confused us, will also unconfuse us.

"I'm sad that we're leaving" is just another way to say, "I'm sad *because* we're leaving." Leaving is the cause of my sadness, not its content. But I cannot substitute "I believe because we're leaving" for my original sentence. Leaving didn't cause my belief. Believing is an intentional state. Sadness is not.

To say that the sensation of seeing turquoise is *about* turquoise is like saying that the sensation of pain is *about* a painful stimulus. A skiing accident might cause a broken leg, but the broken leg isn't *about* skiing. It isn't about anything. We can understand tickling, itching, the pain of a broken leg, the sensation of seeing turquoise independent of any cause. But a belief or desire must be *about* something; one can't be in a state of simply believing nothing in particular. You can't say, "I'm thinking, but not *about* anything." You *can* say, "I'm happy" (period); "I *sense* tickling, turquoise, pain."

> "Fear of what?"
> "I don't know. Fear is fear."
> ". . . Kindness is kindness too."
> (Henry James, *The Awkward Age*)

It's basic to what we mean by "emotion" that we can separate the emotion from its cause. Wistfulness, happiness, sadness, hesitant optimism—these are all states of mind we can ponder on their own, all ways to *be*, regardless of how or why they came about.

Pure Being Achieved?

An important novel throws light on the lower spectrum.

Coetzee's *Life & Times of Michael K* is about a man who just wants to *be*. His intelligence is better than average, but he is often taken for simpleminded because he is uncomfortable with language—the paradigmatic *up-spectrum* way of thinking and communicating. We don't ordinarily *decide* to smile, frown, laugh, blush, or tremble in rage or fear; the body communicates our emotional states implicitly, "automatically," because the body is part of our emotions. But it is separate from our thoughts. To communicate *thoughts* we need a purpose-built system—namely, language, which we invoke explicitly. We must *decide* to speak. Michael K is uncomfortable with numbers and arithmetic too. He shrinks from the upper spectrum. "I am not clever with words," he says. Sometimes he simply won't speak at all.

K's drive to *be* turns all-consuming as he comes to understand himself. "I have never seen anyone as asleep as you are," someone tells him. (Recall von Neumann's being "wider awake than anyone—a radically up-spectrum, versus K's radically down-spectrum, personality.) He lives close to the very bottom of the spectrum, standing beneath an open tap and "turning his face up like a flower" as he is drenched. Put in a hospital against his will, "he lies looking up at the window and the sky . . . , smiling his smile." In Coetzee's words: "Always, when he tried to explain himself to himself, there remained a gap, a hole, a darkness before which his understanding baulked." The bottom of the spectrum is no place for self-awareness. It is a place where being drives out reflection.

K sleeps often, dreams often. He is childlike, presexual, and makes himself a place to sleep nestled on the asexual breast of the South African veld—where "two low hills, like plump breasts, curved towards each other." (Coetzee's symbols are nearly always subtler than this.) K aspires to live like a dream and leave no memories behind, to be translucent, to *half*-exist. "He thought of himself not as something heavy that left tracks behind it, but if anything as a speck upon the surface of the earth too deeply asleep to notice the scratch of ant-feet, the rasp of butterfly-teeth, the tumbling of dust." Yet people such as K "are in touch with things you and I don't understand," says a medical officer. Michael K is a genius at being. "He is not of our world. He lives in a world all his own." As we all do when we dream.

"Back to nature" is the easy, shallow course compared to living low on the spectrum, in the mind's deep end, too saturated with body and being to have mind left over for thought. The novel is about Michael K's decision to live low and *be*. Saul Bellow's heroes are obsessed with something similar. Herzog at the end of his successful struggle: "I am pretty well satisfied to be" (*Herzog*). "You have been summoned," Mr. Sammler tells himself, "to be" (*Mr. Sammler's Planet*).

Thinking and Feeling: Parallel Minds

We assume, in casual thinking, that the mind is created by the brain or the central nervous system. But we all know, on reflection, that there is more to it. Our mental states include thoughts and emotions; mind is created by the brain *and* the body, collaborating. Ronald de Sousa is one of several philosophers who speak of a "two-track mind," the tracks being "intuitive" and "analytic" mental processes. This dichotomy (or something like it) has occurred to nearly everyone who has ever thought about the mind. De Sousa himself points out its relationship to Freud's primary versus secondary mental processes.[4]

We communicate in language, and also body language—which is sometimes as simple as a smile. We can think out a solution or feel our way. "In dance it is not the puppet-master in the head that leads and the body that follows," says a dancer in Coetzee's *Summertime*. "It is the body itself that leads, the body with its soul, its body-soul. Because the body knows! It knows!" We *know* with our bodies, not just our minds. "He knew she was there in the same way he knew when he was in the sun or in the shade" (Philip Roth, *Sabbath's Theater*). We decide and make plans not just logically but emotionally. Churchill wrote in his memoirs about his daring escape from a prison camp during the Boer War. He was behind enemy lines and couldn't think where to turn. "Suddenly without the slightest reason all my doubts disappeared. It was certainly by no process of logic that they were dispelled. I just *felt* clear," he writes (italics mine); he had made the decision in an "unconscious or subconscious manner" (Winston Churchill, *My Early Life*).

Civilized bias in modern times runs strongly for the rational, against the emotional. In normal talk, rational is clean, well lit, tastefully decorated. Virtuous. Emotional is sticky and weak. "He doesn't know what he wants, no focus, too emotional" (Cynthia Ozick, *Foreign Bodies*). Of course, it's also bad to have "no focus." We believe in the upper spectrum. We are warned not to let our emotions run away with us. "That is not yourself, your rational self," says Edmund disapprovingly to Fanny in Jane Austen's *Mansfield Park*. She has been acting on emotional impulses, and he hopes to convert her to rational ones. It is no accident that her emotional impulses turn out to be right—and her rational aunts, uncles, and cousins are all thoroughly wrong. Jane Austen reminds us that there is nothing foolproof about rationality.

Sometimes we think our way to the solution of a problem. Sometimes we feel our way. Sometimes these decisions disagree. We are often in the position of Henry James's Strether, who has decided to send a letter and hurries out, because "if he didn't go before he could think he wouldn't perhaps go at all" (*The Ambassadors*). We're used to having our emotional decisions overruled by our rational, thought-out ones, and sometimes, vice versa. The emotional system is usually much faster, since its conclusions are reached unconsciously. Sometimes we can make its decisions stick by carrying them out before thought has had time to weigh in.

"I know he's guilty" (having deduced the fact) and "I feel he's guilty" (having got that impression somehow) are different routes to the same spot. It takes a writer of Jane Austen's penetration to distinguish the elements of thought from feeling in normal life. Miss Price to Miss Crawford: "'I shall always think of you' said she, 'and feel how very kind you were'" (*Mansfield Park*). In *Persuasion*: "He walked to the window to recollect himself, and feel how he ought to behave." He could *think* it over, but feeling is the better way. Again in *Mansfield Park*: "He did not understand her. He felt that he did not."

Sometimes we can feel our way straight to a conclusion. Feeling in such cases does not play a *part* in thought; it replaces thought. "I never think about these things. I never think and yet when I begin to talk I say the things I have found out in my mind without thinking" (Hemingway, *A Farewell to Arms*).

Physical and mental feelings shade deep into each other. The *physical* sensation of stepping outside into a brisk, beautiful fall day creates the *mental* sensation of (let's say) happiness plus freshness, briskness, anticipation, and a jumble of undertones and recollections. The emotion of *happiness* might cause you to walk a little faster, with more energy; and the physical sensation of *quick, springy walking* might make you (in turn) a little happier. Emotion, hence *mind itself,* requires brain *and* body. Without brain and body, you cannot make a mind.

Sensations are simultaneously mental and physical phenomena, in a sense that does not hold for (say) abstract thoughts. Wordsworth writes of his memories of the river Wye: "I have owed to them / In hours of weariness, sensations sweet, / Felt in the blood, and felt along the heart; / And passing even into my purer mind" ("Tintern Abbey"). Where does physical stop and mental begin? There's no way to draw a dividing line across this continuum.

I might shudder on account of either physical cold *or* a cold, dislikable personality I have just encountered. You hurt his foot, you hurt his feelings—different types of pain, both real.

Body and mind make a feedback loop, positive and negative. You are embarrassed; you blush; feeling yourself blush, your embarrassment increases and your blush deepens. "A smile of pleasure lit his face. Conscious of that smile, [he] shook his head disapprovingly at his own state" (Tolstoy, *War and Peace*).

Thinking and feeling each have their own ways of communicating. When our feelings are strong, our bodies show what we are feeling. Insofar as the body is *part* of the emotion system, it reflects what is going on emotionally. Emotional decisions are made outside conscious control and *expressed* outside conscious control. "He throws up his hands in an odd, unintended gesture. Astonishingly, he is close to tears" (Coetzee, *Slow Man*). The face and body can be a nuanced and precise communication system. Onlookers were "listening for those overtones of the voice, those subtleties of the eyebrows that tell them my true meaning" (Coetzee, *In the Heart of the Country*). We experience emotions bodily and *communicate* them bodily.

Language is a poor way to tell people what we are feeling—and largely unnecessary, because people can tell anyway, if they are half-decent observers. In Jane Austen's *Emma*, Jane Fairfax has "an air of greater happiness than usual—a glow both of complexion and spirits."

Ulysses describes the famously feminine, sensuous Cressida: "There's language in her eye, her cheek, her lip. / Nay, her foot speaks" (Shakespeare, *Troilus and Cressida*). When emotion grabs the body and brandishes it like a flag, the resulting gesture is usually clear—although not always. Sometimes we see only the sheer force of an emotion, driving like a high-voltage wave down to the feet and fingers, out into the universe. The power of emotion, of Guy's horror at the consequence of cheating on his wife, "was enough to make him stop dead in the street and shake his hair with his hands raised and clawlike" (Martin Amis, *London Fields*). Passion can deform or cut off body language, just as it can reshape all other language.

Often we are at a loss to express powerful emotion that beats over the seawall of language and comes crashing, tumbling down into everyday life. King Lear howls like an animal when his beloved daughter is murdered, but howling like an animal is heavily frowned upon; civilized people do not howl like animals, so we are left to improvise. "At the mercy of his grief, with no idea what to do with his misery, he grabbed the janitor's mop, a bucket of water, and a gallon can of disinfectant and swabbed the entire tile floor, profusely sweating while he worked" (Philip Roth, *Nemesis*).

"He said all this very clearly, in a voice without an opinion" (Cynthia Ozick, "Bloodshed"). Words say one thing; the voice might say something else. We converse over two channels at the same time. We control the first. The second, the body-language channel, does what it likes. If words fail, the voice all by itself might get through. Julie understands no Arabic, but "the hoarse flow and guttural hum of the language reached her on a wave-length of meaning other than verbal" (Nadine Gordimer, *The Pickup*). We read tone and we read silence: "'You must miss her.' 'Of course I miss her,' a little too quick" (Thomas Pynchon, *The Bleeding Edge*). We are comprehension virtuosos.

The *mental* aspect of feeling is just one part of a larger picture.

Freud wrote: "He that has eyes to see and ears to hear may convince himself that no mortal can keep a secret. If his lips are silent, he chatters with his finger-tips; betrayal oozes out of him at every pore."[5] Freud discovered that reading his patients' body language was just as important as hearing their words.

"You were sent for," says Hamlet to his sort-of-friends Rosencrantz and Guildenstern, who have been summoned, he quickly figures out, to spy on him: "There is a kind of confession in your looks which your modesties have not craft enough to color." Our bodies, like TV reporters, babble about our emotions ceaselessly. Witness this brief interchange when Ivan Ilyich's brother-in-law has come to visit (Tolstoy, *Death of Ivan Ilyich*): He "looked up at him [at Ivan] for a moment without a word. That stare told Ivan Ilyich everything." Ivan was dying. "His brother-in-law opened his mouth to utter an explanation of surprise but checked himself, and that action confirmed it all."

No words spoken; much meaning conveyed. By imitating in my imagination someone else's precise actions or facial expressions, I will often *feel his feelings* as a result—or, at any rate, guess what they are. Body language flows just as automatically as any other language. In *Ivan Ilyich*, the brother-in-law's actions are involuntary. They are physical expressions of an emotional state.

Emotions get their messages across, one way or another. "He looked very ill; evidently suffering under violent emotions, which he was determined to suppress" (Jane Austen, *Mansfield Park*). Emotion *will* speak, by making you sick if need be. It will speak even if you are "determined to suppress" it.

Thoughts—which have no intensity scale, can't be strong or weak—are *separate* from the body. To communicate *thoughts* we therefore need a purpose-built system—namely, language, which we invoke explicitly.

Brain *and* body are required, again, to make a mind. Today's orthodoxy is never further off course than when it teaches that, since a computer is like a brain, a computer can create a mind. Even if computers *were* like brains, the argument fails, because a brain isn't *sufficient* for a mind. You need a body too. It is a different question whether, if a mind has come into being in the normal way, you can *then* dispense with a body. Maybe so. But no normal mind could have been created in the first place, had the body not been present at the start.

Feeling in the Service of Thought

Tolstoy shows us something important about feeling in the service of thinking. He writes like a blind man feeling his way through life with hands that miss nothing. He was not blind, but he had the supersensitivity of a man who was. This is important because of the nature of consciousness itself. Before Tolstoy writes or even *thinks*, he goes over the ground intimately, half an inch above the surface. Thus we have the incomparable realism of his writing—and a certain awkwardness too.

Emotional decisions are usually instantaneous. Because they are, they can guide us in emergencies when thought is too slow. And because they are instantaneous, it is natural to reach an emotional judgment of everything that passes through consciousness. These emotional judgments (as I will discuss) are powerful summaries, and they make big mental leaps possible; they make possible the linking of two superficially different ideas. Such leaps and links are basic to creative thought.

Emotions are, finally, *inflections of consciousness*. They are colors, tones, varieties of consciousness. If consciousness is bread, emotions are a way to treat or transform it—toast it, cover it in jam, soak it in honey. You can't have "toasted" or "covered in jam" all by itself, without the bread (or something like it). You can't have "happiness" or any other emotion without the underlying fact of consciousness.

A young mother has just lost a child and fears she will forget the fact as she sleeps and will have to face it all over again next morning (a startlingly beautiful piece of writing). But "no, all through her sleep she had known that her child was no longer alive" (Jenny Erpenbeck, *The End of Days*). She knew it by *feeling* it, and feelings inflect consciousness the way colored glass inflects vision. We are conscious when we dream, and consciousness peers through our emotions at the outer and inner worlds.

With this in mind, let's consider, by way of conclusion, Peter Goldie's remarkably concise statement of the intellectual history of emotions:

> There are, on the one hand, those theories that owe their ancestry to the work of William James . . . arguing that emotions are bodily feelings or perceptions of bodily feelings; and, on the other hand, those theories that owe their ancestry to Aristotle and the Stoics, arguing that emotions are cognitive, world-directed intentional states. Other philosophers [including Goldie himself] have argued that, whilst there are analogies to be drawn between emotions and other kinds of mental state, emotions are, at bottom, *sui generis*.[6]

The second idea, that emotions are intentional states, is wrong. The first, however, seems basically right—even though most emotions have no physical expression when one feels them in low-grade, dilute form. When I'm very happy, my heart rate and energy surge. When I'm a little happy, they don't. Certainly, emotions are *sui generis*. But "inflections of consciousness" is a more complete (albeit slightly mysterious) way to describe them.

Summarizing Conscious Mind

Someone hands you a deep-blue wax crayon in a paper wrapper. You experience smoothness, the moderate softness and color of the wax. You experience slickness and the color of the paper, the *look* of the narrow pointed wax cylinder, the waxy smell.

Are you *aware* that you are holding a wax crayon? Not necessarily. If you think about it, yes. But the experience itself—the smoothness, slickness, waxiness—can occupy your mind, and nothing *forces* you to reflect on the experience. Or you might never have seen a crayon before; to you it is merely an object, or a collection of small objects, or a piece of a bigger object. While experiencing something, we are ordinarily, but not *necessarily*, aware of it. Phenomenal consciousness is just experience, pure and simple.

Consciousness can be a nuisance—and not only when you are undergoing surgery.

We all know that certain actions that *seem* to require focused concentration can go all wrong if one becomes "self-conscious." If you are (say) demonstrating an activity for a group, or teaching a class, you can lose the thread if you become *self*-conscious—conscious of your own thought processes. Of the activities you have mastered, some have been mastered by your conscious mind, but many are mastered by the unconscious. Walking is a classic example; we do it automatically, with the unconscious mind in control. Concentrating at the wrong time can be disastrous, because it hands over to the conscious mind control of activities that *should* be controlled by the unconscious.

If you are a tennis player with a blazing first serve, it's your unconscious mind that knows how to do it. If you start *thinking about* how you do it, you will ruin your next serve; you'll have transferred control to your conscious mind, disconnecting the real source of your expertise. If you were finishing up junior high school and wanted to play an infantile trick on a friend who was about to go up on stage and receive an award that *you* should really have got, you would say merely (just as his name has been announced and he is rising to go), "Be careful not to trip on the stairs." As a result, he would be almost guaranteed to trip—because you would have made *walking up the steps onto the stage* a subject for his conscious mind to ponder and take charge of—and probably make a mess of. His *unconscious* mind is the region that actually knows how to walk up steps.

Philip Roth writes about a washed-up actor:

> In the past when he was acting he wasn't thinking about anything. What he did well he did out of instinct. Now he was thinking about everything, and everything spontaneous and vital was killed—he tried to control it with thinking and instead he destroyed it. (*The Humbling*)

Memory, a.k.a. the Unconscious

We have a map of mind, then, with two major regions: consciousness and the unconscious mind. Within the conscious mind, we can make out acting and being, each with its own substructure. But we must also consider the unconscious mind—although the problems here are noncartographic. We don't need further subdivisions, but we do need a clearer view of how things work in this territory.

Our memories are subject to endless distortions, disappearances, misplacements, rearrangements, rewritings. No one believes his memory is flawless. Most of us figure out, too, that we can remember something strongly, positively—and falsely. "Memory plays tricks on us." It's not a deep observation, but it's true and important.

Most of the time, our perceptions are laid out for the mind's consideration in order of arrival, like a crisp line of playing cards laid end to end. If the mind reacts *thoughtfully*, it reacts insofar as *perceptions* trigger *recollections* or *ideas*. We can imagine these *non*perceptions laid out as parts of the same line of cards. As new experiences join the front of the line, slightly older ones are moved from the back of this "very recent" line to the front of long-term memory, memory proper. There, neat chronological sequence comes apart gradually, like the orderly line at an airport gate turning into random chaos (as it always seems to). But some chronological information is retained forever.

Freud knew well that later memories, "screen memories," can distort or corrupt earlier ones. What appears to be a single memory might, in fact, have been stitched together out of several older ones. In such cases we *seem* to remember something that never happened. For my purposes these observations are important but point in a different direction—not to the unreliability of memory, but to the ease with which memory converts a *collection of similar recollections* into a *single template* or schema or abstraction—a process that happens unconsciously but is crucial to conscious thought.

Memory knows only the past but also lets us see into the future—in several ways. Templates, or "schemata," are one of its best tricks. One remembers the recent past as a narrative outline, a temporal sequence. Aside from this "retentional" aspect of memory, there is also (Husserl noticed) a "protentional" aspect: we can see what will happen next. When life has been following a series of steps, as it almost always does, we can usually see the next one clearly enough that we won't stumble. This method of foreseeing the future is short-range but important. A cadence that has paused on the dominant will resolve to the tonic; the harmony won't be left hanging. A bird that has stroked and glided halfway across the sky will keep going and cover the second half too. The turning-over, throat-clearing engine will thrum to life.

We make predictions over longer timescales too, and we learn to make new ones as new *predictable patterns* crop up. Now that I have handed money to the cashier in the little booth in the parking garage, my change will be handed back out. Memory's tasks go far beyond supplying reminiscences and facts. Memory is a pattern recognizer, discovering and supplying us with the knowledge of patterns we need in order to get through the day. I have sent a file to the printer and will soon hear the printer start to work. We have sat down at a table, so a waiter will materialize and take our order. Such schemata work the same whether they cover a short timescale or a long one—whether I have in mind a fast-running template for flattening a mosquito or a slow-running template for driving from New York to Washington.

Occasionally, we are taught such schemata. Usually, we discover them for ourselves; to be more precise, memory discovers them for us. These temporal schemata are the same as spatial ones: we know how a typical school day is organized, and we also know how a typical school building is organized. We know the template or schema of a typical suburban split-level, or a table set for an informal dinner, or a southern-English medieval cathedral, or a fast-food hamburger. We know these just as we know the typical architecture of a visit to a movie theater or a dentist, or a TV newscast, or the progress of a presidential election season—an arrangement of elements in time.

Some mind researchers insist that temporal and spatial schemata are two different topics, to be understood separately. We have always found it irksome, somehow, to recognize one pattern across space *and* time.

Midlength temporal schemata, such as visits to the dentist or a restaurant, are sometimes broken out as a special category just in themselves; they are called "frames" or "scripts." But along with the paradigm maple tree, or peanut-butter-and-jelly sandwich (in the National Bureau of Standards), they are all just schemata, just templates.

We discover temporal patterns and spatial ones the same way. *We learn by forgetting.* When many separate memories are largely the same, we tend to forget the little differences and blend those memories together into one abstract, heavyweight memory. The result is an *idealization,* based not on analysis but on experience. David Foulkes writes: "The second time I stay at a particular hotel I may be stimulated to think back to the first, but by the tenth time I may simply have general knowledge of the hotel rather than much vivid recollection of any one past set of experiences there."[7]

Such blending (such melting together) is especially apt to happen when we are children. Whether over space or over time, the process is the same. Once I have seen many apples, my individual apple memories blend together and give me a template or schema for a generic apple. It's red, or maybe yellow or apple-green, roundish, with a stem on top and a pucker at the bottom, and it crunches when you bite it. *Time blending and space blending work the same way.* A schema is a schema.

Feeling and Memory

Feeling is a creature of the conscious mind. Memory is the country next door. But we cannot understand memory unless we understand the role played by feeling in making memory work.

Up-spectrum, we create abstractions by focusing on one aspect of *many* recollections. For example, think about color in hundreds of apple memories. Down-spectrum, we tend to focus on one recollected episode at a time—the whole thing, not just a piece of it. Just as one aspect of many recollections yields an abstraction, *all* aspects of one recollection (or of one experience as it actually happens) can yield an *emotion.*

Some experiences are compact and simple. Look at your watch;

decide you're late; get going. But some have many parts. The difference lies mainly in how you choose to experience each event—the pace at which you're moving, your position on the spectrum. A multipart experience might be the forest shrugging off a warm, mild breeze with a faint rustle of leaves like dim applause, and a shuffling of sun prisms on the ground. Add the smell of moist bark and moss and leaf-laden soil, busy flies, a distant barking, and the soft humid air, the tentative trickle of a drowsy summer brook; heat; mist; stillness. But that whole scene feels *one* particular way; some leafy, summery, shadowy emotion without a name is how it makes you feel. The whole scene creates *one* particular mood. One of Cynthia Ozick's characters aspires to compose a "dictionary of feelings"—"feelings that everyone's somehow felt, only there's no name for them" (*Foreign Bodies*).

If someone asked as you stood in the warm forest, "How do you feel?" you might answer, "sweaty, peaceful, moody," but the real emotion is a single feeling, and more subtle than those three words suggest. That feeling or mood is a *summary* of the one scene—just as a template or abstraction (of an apple, a tree, a Manhattanite) summarizes one aspect (or several) of *many* recollections. This sort of emotion summary can bait a hook that lets you fish one particular recollection out of the deep ocean of memory. You can't recall a memory with nothing to go on. Something must lead you to it—some clue, some fragment of the scene, some association. At low focus, emotion summaries lead us to memories that would otherwise have been lost. Without emotions as bait, we would have been left with no way to recover them.

The role of emotions as summaries or abstracts is a deep, fundamental fact about the mind. There will be more to say about it, when we look at the spectrum zone that encourages this function's emergence.

For now, "her memory of her father came to mind as we were talking about her argument with Mark." This is the psychoanalyst

Stephen Grosz describing a patient. "The two events *felt* similar to her."[8] Why did the first memory suggest the second? A shared *emotion* was the connector, the memory cue. They "*felt* similar." An emotion connected one memory to the other.

The psychologist Endel Tulving laid down, in a seminal work (*Elements of Episodic Memory*), the "encoding specificity principle," now widely accepted.[9] "Encoding" means creating a memory; installing a particular scene or episode in long-term storage. You are more likely to recall something if you reproduce *the environment in which the memory was created*. If you are in just the same place, or just the same mood, you are re-creating some part of the original environment. And you are more likely than usual to recall things you first memorized in the same circumstance. The environment in which you memorized something is apt to become part of the memory, so *revisiting* that environment creates a powerful memory cue. The environment, which forms part of the memory, naturally tends to bring the rest to mind.

You can look at a person and recall that you have met him or her before. (Berowne to Rosaline: "Did not I dance with you in Brabant once?" Rosaline, in response: "Did not I dance with *you* in Brabant once?" [Shakespeare, *Love's Labours Lost*].)

You can travel to a particular place and recall things associated with that place. On arriving at a hotel in Rome where I had stayed before, I recollected a market next door where you could buy bottled water but not Tylenol. This fact couldn't have come to mind under any other circumstances.

This is a good example of Tulving's "encoding specificity"; it's also a good example of the mind's ability to distill the many elements of one memory into *one* summary mood—just as it can reduce one element of many memories to a template or abstraction. It wasn't the golden-yellow stuccoed façade or the small, darkish lobby of the hotel, or the room or the furniture or framed prints on the walls, or any other single feature that brought the place next door to mind. It was the whole package.

Software shows us how difficult it is to search for something given a long list of criteria—façade this color, walls that color, rooms like this, furniture like that—instead of just one. (What's Dumbo Schwartz's real first name?) The best way to proceed is by computing a single code based on *all* those criteria, often called a "hash code"; and that's what memory probably did in this case, as it does in so many others. All these details together yield one particular ambience or mood or feeling. That feeling is a search key that allows us to find other memories marked with that same mood.

Consciousness can distill and synthesize. This is one of the mind's most important powers—a central facet of memory and recollection.

Thought and Memory:
The Map from Overhead

Conscious mind and memory each have their separate functions, but the way they *collaborate* is essential to the whole works.

We dip into memory constantly as we think—as we tap the space bar when we type. Reference to memory is *part* of thinking. Intercourse between perception and memory is continual, and it grows in volume as we descend the spectrum. We see someone and think, after a quick memory check: "He looks happier than usual" or "older." We feel a back pain and think "not again," hear music and think "isn't that . . . ?", see a tree and think "it's early for dogwood blossoms." "Most researchers," writes Jerome Singer, "now believe in general that the act of perceiving external stimuli cannot occur effectively without some matching of this stimulus with material from long-term memory."[10]

But how does recollection *work?* An external event *or* a thought can make us remember—can serve as a "memory cue." We might recall a day at the races because we hear a bugle call, or because we have had one described to us, or read about one, or otherwise come to think about a bugle call.

Any thought or thought fragment can bait the hook that fetches recollections out of memory, can be a "memory cue" or "retrieval cue."

There is no difference between a memory *cue* and a memory. Each is a snippet of experience. Each can be roughly approximated, for convenience, by a list of features: forest, noon, sunny, humid, flies droning.

Free association can be shallow or deep. The shallow variety can often lead to something deeper. Superficially, I recall a movie I saw long ago in a theater near the Plaza in Manhattan, then I recall something else that happened near the Plaza or in it—or something associated with the movie, or with Paul Scofield, an actor who was in the movie.

But this sort of superficial ramble often leads to deeper free association—in fact, to *theme-circling* thought. In this deeper free association, one memory *as a whole* suggests the next.

The theme might be trips to Manhattan when I was in high school, or something deeper—slipping into forbidden R- or X-rated movies with friends when I was in high school, and other banned activities, or something else entirely. The theme isn't stated explicitly. It is implied by the thematically related sequence of thoughts. Yet despite being implicit, the theme is stated *plainly*, in the sense that a circle shows us its center point plainly, even if the point isn't marked.

Theme-circling thought, then, is a series of related recollections, once we have sunk a fair distance down-spectrum and conscious mind has relaxed its grip and let memory go off on its own.

In Sum

We have the tide chart and the map. With both in hand, we can go out and tour the spectrum, starting up in the hills at the high end (the spectrum has now become a brook in New England) and continuing alongside rapids and waterfalls until we arrive at the deep water and the past at the bottom.

Five

Spectrum, Upper Third: Abstraction

In the upper spectrum, the conscious mind runs the show and memory is subordinate, a tool to be used and controlled by consciousness. Conscious mind takes action; this is the realm of *doing*, not being. How do we decide what needs our attention? Usually we just follow routine. Sometimes a pressing problem brings us up short: emotion calls. (Forgot I need gas. Do I have time? Anxiety pang.) But often we make decisions the easy way: whatever shows up first gets our attention. Usually we barely notice ourselves choosing.

In the upper spectrum, emotion is important in *getting* our attention ("our" meaning the *conscious mind's*). Otherwise emotion is, mainly, excluded.

The mind is made of *consciousness* (thought and feeling) and *memory* (also known as the unconscious). In the upper spectrum, thought is in charge and we are free to reason. Reasoning is usually our best route to resolving an unfamiliar problem or deciding on a proper response. Before we plunge into reasoning, though, we make a routine check: can I recall the solution to this problem *without* thinking? just by remembering?

If the answer is no, we will think out our problems or plans rationally. "Rationally" implies logically or systematically and, as far as possible, by the use of abstraction. In other words, we will not be caught in irrelevant detail; we will get straight to the point, right to the heart of the matter.

Abstraction yields *conciseness*. (The "abstract" in a scholarly paper is the synopsis at the start.) Abstraction means skipping detail and special cases. High analytic intelligence, high IQ, makes you *quick*. You are quick because you wield abstractions confidently and use them at the highest level—use the most *abstract* abstraction that seems promising. Abstraction is the defining procedure of the rational mind.

In rational thought as in all thought, we depend on constant queries to memory. In thinking of all sorts, memory queries are like breathing. In the rational upper third of the spectrum, the conscious mind's relation to memory is simple and well-defined. The conscious mind is in charge and uses memory as a tool. Memory is kept on a short leash and is not allowed to wander. The conscious mind makes focused, specific queries to memory and gets information back. Not reminiscences; not anecdotes. Just information.

We often think of memory as a warehouse of separately packaged recollections. It is, but it is also a tap. Turn it on; information flows. Memory functions in the upper spectrum mainly as a giant experience juicer that squeezes memories for the data they contain. It is a computer (or better, a wise reference librarian) that answers questions (how do I get a large log out of the road?) with information. "Get many people to lift, or chain it to a car or truck and drag, or cut it up small." Sometimes the data is fresh-squeezed. Sometimes you can recycle the answer from an earlier occasion and don't need to resqueeze it.

Memory uses the recollections it stores as fuel to satisfy requests. Memory is, among other things, a great bin of oranges, grapefruits, and pineapples awaiting juice orders. Orange juice is *itself* a synopsis, after all. It's the essence of a bunch of oranges. It's an abstract.

Behind the scenes, memory does its own abstracting unconsciously; sometimes concurrently with its conscious assignments.

This unconscious activity happens all by itself. No entity is required to give the order that it be started. *Templates*, as I have said, are crucial to thought: a template, or schema, is an *abstract* of a bunch of related recollections. Now that we are discussing memory as an essence squeezer, as an "abstraction engine," we need to look at templates, or schemata, again.

What is a tree? What is a New England forest like, or a successful Parisian businesswoman? These requests are about *space*, abstractly speaking. A tree is an object with a certain structure. If we are asking about the Parisienne's appearance or the impression she makes, we are asking again about a certain spatial *structure*. But if we are asking what her life is like, or her typical day, we are asking about *time*. Templates provide answers to both types of questions, spatial and temporal.

When it creates templates, memory *learns by forgetting*: by strengthening or underlining points that are true for most maple trees (say) and blurring out, or forgetting, details that are atypical. It is a simple, powerful process, beautiful in its simplicity and generality. Beautiful in the vast variety of things it can do, this one simple operation. It can make a template for *any* sort of object, in space or time (an "object in time" being merely the sequence or narrative, the "object" whose parts are arranged in time—the visit to the dentist or gas station, using the ATM). And the same simple operation allows you to forget details that are merely cluttering things up.

Memory must conserve space and protect efficiency by sweeping away clutter. It must conserve space and improve performance by creating templates—so that it can make good guesses, anticipate what's coming, guide the boss's behavior so he doesn't make a fool of himself—if possible. Memory must collapse, compress, meld eighteen similar gas station memories or five Parisian businesswomen into a single template, a concise guide. The exact same operation that gets rid of clutter.

The template creation I'll describe is crucial to the mind's functioning—and is wholly passive. The mind doesn't *do* anything—in the sense that adding numbers, searching a poem in your mind, or reading a book are mental *doings*. Template creation is the mere natural "settling" of separate recollections, the way rocks and soil settle. If you heap a pile of soil or leaves or mulch in your backyard, it will settle—will compact itself. A heap of similar recollections undergoes a similar natural process.

The psychologist and memory specialist Ulric Neisser writes, "Increased experience with any particular event class increases semantic (or general) knowledge about the event and its context. Increased experience with similar events, however, makes specific episodic knowledge increasingly confusable, and ultimately episodes cannot be distinguished."[1] In other words, ultimately we compact similar episodes, losing detail but gaining a general guide.

We can also *invoke* this ordinarily passive process of settling on purpose, consciously, when we are asked for general knowledge, for what to expect, for the wisdom of experience.

Making Templates

Suppose you face an everyday problem: The car barely turns over and won't start. You can't think of a present for Olivia. Someone hasn't turned up for lunch and won't answer the phone. You left your coat at the doctor's office.

What do you do? You might recall an earlier experience and do again what you did then. Or you might recall many earlier experiences and turn them into a template on the spot, simply by superimposing or conflating them.

How do you conflate or compress or "focus" a collection of similar memories? You meld them together, letting frequently occurring aspects of separate memories reinforce each other, and seldom-occurring aspects cancel each other out.

Imagine superimposing a dozen translucent photos—twelve quick snapshots, let's say, of one fashion model. If her face shows up in the exact same position, with the same expression, in all twelve (and they are aligned), then the stack of superimposed photos gives a clear image of the face. If her right arm is posed differently in every picture, you get a blur instead of a right arm. If her left arm is straight up in nine photos and straight out in three others, you get a straight-up arm that is stronger and brighter than a fainter, straight-out image of the same arm.

Notice that you've lost information in creating the template. You no longer know whether the photo in which the right arm is pointed forward, say, included a straight-up or straight-out left arm. But ordinarily, you don't care. Your up-spectrum mind conflates twelve stuck-car memories in roughly this same way. Elements that are nearly the same in each memory are reinforced and stand out in the conflated memory.

Let's say your car won't start and you're conflating twelve separate recollections of similar events. (Of course, you aren't aware of the twelve separate memories or the conflation process itself; you are merely *remembering*.) "The car barely turns over and won't start" is part of each recollection. That's why they were all summoned to begin with. It stands out. "Dead battery" is part of many of the same recollections, maybe all, and *it* stands out. But the location where the car got stuck, the type of car, the time of day, the year, the weather, and the occupant of the passenger's seat will vary. Those details probably blur out and disappear. It *could* be that nine of the twelve memories all deal with the same lovable, treacherous Audi you used to drive. In that case, "Audi" stands out almost as clearly in the melded-together template as "dead battery." But that's just an artifact of the data, an accidental detail. You edit it out "by hand," deliberately, because you know that stuck cars are not, *in general*, Audis.

This conflated, compressed supermemory is exactly a *template, schema,* or *abstraction.* It is a memory sandwich, a "heavy-duty memory" with two main elements that emerge clearly: "won't turn over" and "dead battery."

The more separate memories become compressed or melded together, the more likely it is that everything will drop out or blur out *except* those elements that really do go together, that are essential and not accidental to the template.[2]

Learning by Forgetting

You might think similarly about (say) trees: remember many separate instances and conflate them. The result would be a compressed, conflated, melded supermemory, *the memory of a nonexistent object*, of an abstract or paradigm tree. This abstraction, template, or schema might consist of *trunk, branches, green* (represented visually, not in language). Let's say two-thirds of your original memories had leaves and one-third, evergreen needles. If so, your tree abstraction has *leaves* clearly defined, *needles* less clear but plainly present. All other details are blurred out. And that's your tree, abstractly—the bare bones. It has trunk, branches, greenness; often leaves, sometimes needles.

Usually such abstractions are created behind the scenes as part of routine, unconscious mental housekeeping (as I have said) *when you are a child.*

"Objects, on our first acquaintance with them," writes the nineteenth-century essayist and painter William Hazlitt, "have that singleness and integrity of impression that it seems as if nothing could destroy or obliterate them, so firmly are they stamped and riveted on the brain" ("On the Feeling of Immortality in Youth"). But as memories are added on top of memories, settling and compression are natural. If I saw my first maple on Monday, my second on Tuesday, and today is Wednesday, the interval separating the two sightings is half the age of the oldest. So the two sightings do not blend together; their ages are very different. A month later, the same two original sightings are separated by only one-thirtieth the age of the oldest. Now their ages are similar and no longer prevent their blending.

It's natural for a child to forget (unconsciously) the distinction between one tree memory and another. It's natural for those memories to blend together into a single conflated, abstract "tree"—which continues to develop as more individual memories are added to the melted-down abstraction. Only tree memories that stand out in some striking way remain distinct.[3]

To put it differently, as you accumulate similar memories, you tend to confuse them. Confusion appears in two ways. First, you can no longer distinguish separate episodes; you can no longer remember that you saw a large maple tree in a park last Sunday and a small one beside a house three weeks ago. Your "episodic memory" is failing. On the other hand—second—you can make assertions about maple trees *in general* and feel sure about them, without thinking about any particular example. "Semantic" memory is emerging as "episodic" memory fails—*semantic* memory being your store of general facts, rules, principles, expectations.

Melding memories so that common features emerge and individual details—atmospheric idiosyncrasies—disappear makes high-focus thought powerful, and numb.

So we learn by forgetting. We learn what "tree" means by gradually, when we are young, forgetting the differences among separate trees and remembering the common points.

Event templates work the same way. We have pulled into a gas station many times and know the routine. This template will be a timeline. Again it's convenient to imagine a memory or template as a filmstrip—in time instead of space. Each image is later than the one before. The template is created the same way as any other. We accumulate many memories of buying gas and they settle; they meld together. Extraneous elements blur out; those that occur every time emerge brightly.

A "timeline" or "event" memory is no different from an "object" or "scene" memory. Memories are quotations from reality and can be read as sequences. You pull up to a pump, get out of the car, swipe your credit card, and choose your gas grade, or you wait for an attendant and say what you need. Such templates can also be read as static images. What is a gas station? It's a concrete-paved rectangle with pumps on an island.

As I've mentioned, some psychologists insist that templates for events—often called frames or scripts—are different from templates for objects (or templates for ideas). But if we are serious about psychology, *Occam's razor applies*: do not posit two mechanisms or three or a thousand when one will do. *Pluralitas non est ponenda sine necessitate*; plurality is not to be posited without necessity.

Template creation—amalgamating separate memories so that aspects shared among many are emphasized, and unshared attributes are de-emphasized—is a neat and elegant operation. It accounts for the creation of templates for objects in space and in time. It accounts for their gradual, incremental accumulation over years, as new individual instances join existing templates and blend in.

Those individual instances are *attracted* to the existing template by exactly the same like-attracts-like principle that governs all of memory. Someone asks the name of a tree, and I say it is a peach tree because the attributes I observe *call to mind*—like attracts like—the memory of a particular peach tree *or* a peach tree template. By just the same process, when I see a maple tree and memory records it, that new maple tree recollection is attracted by shared characteristics to the existing maple tree template. *Like attracts like.*

The whole memory runs on like-attracts-like plus natural settling plus momentum. Like-attracts-like handles queries ("tell me about orange juice") by marshalling memories that include, and therefore match, "orange juice." Like-attracts-like means that similar memories about anything tend to cluster together. Natural settling means that two memories that are closer than a particular threshold melt together. Momentum deals with all the transitions from a well-behaved, docile memory answering information requests to an off-on-its-own, free-associating recollection surfer following the "isofeels," following one feeling from memory to memory.

When this sort of surfing turns up a blocked emotion as part of an ordinary recollection, the emotion tends to take over. Blocked emotions are powerful things—as an unsatisfied urge always is; as a natural action cut short, sliced in half, left suspended in limbo, always is. When William Blake wrote, "Sooner murder an infant in its cradle than nurse unacted desires," this is, I think, what he meant. The power of a *blocked* natural action is gigantic.

The philosopher Georges Rey is right in saying that "there is every reason to think that human beings are not ideally designed, but are a hodgepodge of some very arbitrary evolutionary accidents."[4] Fair enough. Anyone can think of aspects of the human creature he would love to see improved. (*Immediately.*) But at the same time, Rey and many other mainstream mind thinkers give us the feeling that they don't quite see the beauty of the mind we have. It *is* buggy and fragile, and subject to grotesque abuse (as any investigation into good and evil or into freewill reveals). It is delicate, absurdly sensitive—a far more sophisticated design than really made sense under the circumstances. It was a splurge that has gone wrong in the field again and again, in a million ways. Still: how beautiful.

Enter Reasoning

How do we assemble thought trains when we reason? What determines which thought follows which other? How do we construct a train of thought?

"Train of thought" does not mean that our thoughts are laid out neatly end to end like dominoes. Often thoughts blend into one another, like colors in a tube of light fading gradually from one to the next. Sometimes we think of nothing and merely register external stimuli. ("What are you thinking?" "No thoughts. Isn't that nice?" "It's sublime" [Philip Roth, *Deception*].) Sometimes we get stuck and the mind spins its wheels. But "train of thought" is a useful idea if we know its limitations.[5]

Simple problem solving is a place to start. Suppose you notice that you have lost your keys. They're not where you ordinarily keep them. You need them.

Wherever you are on the spectrum, it is always simpler to remember a solution than construct one. You aren't *obliged* to perform careful reasoning when you are up-spectrum; you are merely capable of it. When you notice the missing keys, you might recall that the same thing happened last week—and you had let your son take the car the day before, and asked him to leave the keys on the kitchen table when he got back, and that's where you found them. And he took the car last night too. So you look on the kitchen table, and there they are.

But suppose they're *not*. Suppose memory has no easy answer. You have to think. You switch on your powerful rational thought engine. Scanning a timeline seems like a good first step. You drove home yesterday evening, took the keys out of the car, and put them, presumably, in your pocket. You follow the line forward: what could have happened next?

You are supposed to be reasoning, but here we are talking *time-lines*, not *modus ponens*, not inference. Why?

Rational thought is more than logic; more than gathering all your data and assumptions and pushing your way forward. Rational thought also means picking out the right techniques and the right shortcuts—the right heuristics. Following timelines is a crucial heuristic.

Timelines are also close relations of logic. Reasoning implies an abstract timeline. *Before* I can split chunks, I need to cut up the log. Before I can cut up the log, I must find a chainsaw.

Causation in time is not the same as logical implication. *Having a chainsaw* implies that *I can cut up a log*. Here we say nothing about time. We say merely that the truth of the first proposition guarantees the truth of the second. But there is a natural path in the woods between the timeless, elevated, pure road of logic and the more convenient path (winding alongside the stream) of timelines and causation.

Timeline thought is close to abstract reasoning. The consistent time thread that runs through our entire lives is a help in problem solving. I do what is reasonable, and reason transcends logic.

Mentally you retrace your steps, then, from the moment you left the car last night. What could make the keys disappear? Hole in your pocket? Lent them to someone? You consider each possibility. This is rational, systematic thought—unemotional, disciplined.

This sort of thought requires sustained attention. We tend to lay out systematic arguments step by step like garden flagstones as we build a path from premises to goals. But we rarely fill in the path sequentially. Often we work backward; often we cover important points along the way and then fill in the gaps.

Reasoning no less than reminiscing depends on a series of recollections. It is driven forward by requests to memory satisfied one by one. Consider some small examples.

In dire circumstances, Viola in Shakespeare's *Twelfth Night*, unluckily shipwrecked on the shores of Illyria, makes a plan to survive. Shakespeare has no interest in illustrating rational thought per se. But he gives us a good example of emergency systematic thought: "I pray thee"—to the captain with whom she is shipwrecked—

> and I'll pay thee bounteously—
> Conceal me what I am, and be my aid
> For such disguise as haply shall become
> The form of my intent. I'll serve this duke.
> Thou shalt present me as an eunuch to him.
> It may be worth your pains, for I can sing
> And speak to him in many sorts of music,
> That will allow me very worth his service.

In other words, *I'll pay you to do this: conceal my identity. Help me disguise myself. I'll support myself as a servant to the local duke. You've said he has banned women from his court, but no disguise will make me look like a man. So, you'll say I'm a eunuch. And I'll succeed in getting a job at court, because I can sing. In fact, I'm a versatile musician.*

The goal? Survive. How? Get a job. Working at court for the duke is the best job around. How to get such a job? (1) Offer useful services, and (2) be male. So we've broken the problem down into two subgoals.

What services? Easy: I'm a musician.

How to be male? Disguise; but no disguise will make me look like a man. *Dead end.* Can I think of a different way to move forward?

So far, we have seen systematic working backward from the goal to the step right before, and the one before that. The process is accompanied by frequent requests to memory for information. What does a duke need, and is there anything on that list *I* can supply? Memory provides the data. Then how can I seem like a man? Memory responds. You can wear a disguise; *but*, if you are in close contact with the fellow, forget it. No disguise will work.

Memory requests are incessant in rational thought. Bailing out from abstract propositions to images is probably incessant too, although that depends on the problem *and* the problem solver. Probably, Viola imagines herself—*pictures* herself—in male disguise, in the duke's presence. She doesn't conclude *logically* that it can't work; she *sees* that it can't work.[6]

We are at a dead end. Maybe there is some give in the requirements? Having in mind, simultaneously, how *can* I disguise myself and what *might* the Duke accept, "eunuch" pops out.

Memory operating up-spectrum, in juice-squeezer, information-please mode, together with simple rules of logic and basic life experience, produces an impeccable solution. Further down-spectrum, Viola's train of thought might have wandered off into sad recollections of better times, or reflections on what makes a good disguise, or an excursus on courtiers she has known. But up-spectrum, thinking is focused, disciplined, and systematic.

In *Anna Karenina* (Garnett translation), Lyovin's wife is about to give birth, and he is "conscious of an increasing uplift of his physical powers and of his attention to all that lay before him." He makes his plan. "Kuzma should go with a note to another doctor, while he himself would go to the chemist for the opium; and if the doctor was not up when he returned he would bribe the footman—or if that was impossible, he would enter by force and wake the doctor at all costs."

Again, the problem is broken down repeatedly. Lyovin rises to the occasion; notice *how* he does. When you face emergencies, your body *moves* you up-spectrum, like a chess piece. He *feels* the surge of energy, in his physical powers—no doubt in his mind too. He feels *focus*: his mind is now dealing with one problem alone, concentrating all its energy on his wife. He works his way systematically backward from goal to starting point and, having laid out a path, sets out.

Not everyone is capable of it, or willing to bother. Those who are willing and able are most likely to do it when they are wide-awake, energetic but calm—*up-spectrum*.

High-focus thought will not be blown away or pushed off course by an interesting but irrelevant perception, recollection, idea, or (especially) *emotion*. The high-focus thinker is deliberately "narrow-minded"—the mind is focused (is *un*dilate) and, like an eye's narrowed pupil, admits a minimum of external glitter and glare.

Exit Reasoning

It hardly needs saying that there is far more to reasoning and abstraction, to the upper spectrum, than I have said here. But these have always ranked among philosophy's favorite topics. The mental phenomena of the middle and lower thirds of the spectrum are the ones that suffer from inattention. They are not only intuitive and unreasonable—or at least *not* reasonable. They tend also to be concrete rather than abstract, visual rather than language based, subjective rather than objective. For all these reasons, the middle and lower spectrum thirds put modern thinkers on edge. They are a bad match to our science-venerating intellectual climate, and (in many, if not all, respects) to the modern philosophical tradition that dates to Descartes and the proto-modern tradition that starts with Plato.

These middle- and lower-spectrum phenomena—the ones I will discuss in the next two chapters—are also the ones that pose the hardest problems if we are bent on computer minds. It seemed reasonable from the start that computers could be used to simulate or reproduce logical or rational thought. But no one ever said, faced with a first-generation "electronic brain," "I'll bet I can make it happy. I know a joke it will like." There was no reason at all to imagine that these machines were *capable* of being happy. Why would anyone make a machine that was capable of feeling? Or consciousness? Even if you knew how to do it, what would be the point?

In his famous 1950 paper about artificial intelligence, Alan Turing mentions consciousness, in passing, as a phenomenon associated with minds, in some ways mysterious. But he treats it as irrelevant. If you define the purpose of mind as rational thought, then consciousness certainly *seems* irrelevant. And for Turing, rational thought was indeed the purpose of mind.

Turing's favorite word in this connection is "intelligence": he saw the goal of technology not as an artificial mind (with all its unnecessary emotions, reminiscences, fascinating sensations, and upsetting nightmares), but as artificial *intelligence*, which is why the field has the name it does.

In no sense did this focus reflect narrowness or lack of imagination on Turing's part. Few more imaginative men have ever lived. But he needed digital computers for practical purposes. *Post*-Turing thinkers decided that brains were organic computers, that computation was a perfect model of what minds do, that minds can be built out of software, and that mind relates to brain as software relates to computer—the most important, most influential and (intellectually) most destructive analogy in the last hundred years (the last hundred at least).

Turing writes in his 1950 paper that, with time and thought, one might well be able to build a digital computer that could "enjoy" strawberries and cream. But, he adds, don't hold your breadth. Such a project would be "idiotic"—so why should science bother? In practical terms, he has a point.

To understand the mind, we must go over the ground *beyond* logic as carefully as we study logic and reasoning. That's not to say that rational thought does *not* underlie man's greatest intellectual achievements. Cynthia Ozick reminds us, furthermore, of a rational person's surprise at "how feeling could be so improbably distant from knowing" (*Foreign Bodies*). It's much easier to *feel* something is right than to prove it. And when you do try to prove it, you might easily discover that despite your perfectly decided, rock-solid feeling of certainty, your feelings are total nonsense.

We have taken this particular walk, from the front door to the far end of Rationality Park, every day for the last two thousand years. Why not go a little farther this time, and venture beyond the merely rational?

Six

Spectrum, Middle Third: Creativity

How does creativity work? Few questions in all of mind science, or in science, philosophy, and psychology in general, have been asked with such keen interest in modern times. We worship creativity, and we know it is rare. If only we could *understand* it; then we could teach it. And then *everyone* would be creative! This simple and ultimately inspiring, even moving, thoroughly unconvincing belief survives from the progressive age—the generation after the Second World War. In those years, Americans were sure that anyone could learn anything.

Creativity is a hard problem because there is no step-by-step way to achieve it. We can learn how to solve elaborate logical or mathematical problems, learn how to translate foreign languages or play baseball or fly a kite or drive a car. To do any of these things in an inspired way requires special gifts, but nearly anyone who tries can reach basic competence. That's what's perplexing and frustrating about creativity. There is no way to reach even basic competence.

We *can* understand creativity, or a good deal about it. Creativity has much to do with the dynamics of the spectrum and two of the spectrum's major transitions: the gradual emergence of emotion, and the unconscious mind's gradual taking over from consciousness, as we move down-spectrum.

Diffuse Attention Is Required

The physicist and philosopher Roger Penrose writes that creative thoughts are apt to come to him as he thinks "perhaps vaguely" about a problem—"consciously, but maybe at a low level just at the back of my mind. It might well be that I am engaged in some other rather relaxing activity; shaving would be a good example."[1]

Penrose mentions, too, the great mathematician Jules Henri Poincaré, who found the key to a hard problem while getting on a bus. "This complicated and profound idea apparently came to Poincaré in a flash, while his conscious thoughts seemed to be quite elsewhere."

According to the literary philosopher and novelist George Steiner: "All of us have experienced twilit, penumbral moods of diffuse attention and unresistant receptivity on the one hand, and of tensed, heightened focus on the other."[2]

Thus, *Penrose* is creative when he is thinking "perhaps vaguely" about a problem—"consciously," but just barely.

Poincaré finds a creative solution to a problem "while his conscious thoughts seemed to be quite elsewhere." He must be barely conscious of the problem, or not conscious of it at all.

Steiner awaits inspiration—he experiences "unresistant receptivity" —during "twilit, penumbral moods" when his attention is "diffuse."

In other words, creative solutions arise when a problem is lurking at the edge of consciousness. Logical solutions require focused attention (attention dialed up, metaphorically, to a bright, sharp spotlight), but creative solutions arise at a much lower level of focus—attention, metaphorically, creating a broad pool of light. Penrose uses the phrase "low level" and Steiner writes "twilit, penumbral"; both thinkers have used metaphors that fit the spectrum perfectly. Creativity occurs (in these cases anyway) when focus is fairly low. We are not at the bottom of the spectrum; if we were, our thinkers would be on the way to sleep and would probably not be self-aware and reflecting on their mental states. So, "fairly low focus" seems reasonable.

The Psychological Profile of Creativity

Creative inspiration is a mental event that has been discussed at length; it has a well-defined profile. We know what to expect and can make inferences from our expectations.

You don't creep up methodically to a creative insight. It hits you on the head unexpectedly. You can make yourself work hard on a problem in algebra and solve it step by step. But you cannot make yourself be inspired, cannot make yourself experience a creative insight. In this respect, creativity is like falling asleep. We can encourage sleep or creativity by setting up the proper environment. But we cannot make either one happen (short of using drugs in the case of sleep).

Creative inspiration does *not* give you a feeling that you are approaching a solution, "getting warmer" as you work the problem. An intriguing experiment by Metcalfe and Weibe showed that test subjects searching for a creative, out-of-the-blue solution to a set problem *felt* no closer to succeeding just moments before inspiration struck than they had felt at the very start.[3] When subjects solved problems calling merely for correct application of rules, the opposite held. They reported feeling closer, "warmer," as they approached a correct solution.

Creative inspirations feel as if they hit us out of the blue. They are unprovoked slaps in the face. The great classicist E. R. Dodds writes that inspirations "come suddenly, as we say, 'into a man's head.'"[4] What does this mean in spectrum terms? Ideas hit us out of the blue when we have started to lose control over our own thoughts—that is, we are moving lower in the spectrum. But we cannot have gone so low that we fail to notice what is happening. We must be able to notice that we have scored an inspiration, and be able to do something about it—or at least remember it until we are more focused.

Analogies Are Basic

Creative problem solving is widely agreed to center often on *inventing a new analogy*—sometimes called "restructuring" the problem. When you suddenly see a connection between two things you don't ordinarily speak or think of together, you have the basis of a new analogy, or a creative thought. *That branch on the ground is like an arm and a reaching hand.* Whereupon you might pick up the branch and retrieve something that was out of reach before.

Most inspiration happens at this shrug-it-off level of ordinary, everyday, unremarkable thought. But many important inspirations have the same modest beginnings: *new analogy*. By comparing a puzzling *something* with a *something else* to which it's never ordinarily compared, you tear an opening in the everyday fabric of mental life and peer through. You can now think about the something in terms of the something else. You can think about it in a new way.

In this view—not a view connected to the spectrum, merely a standard idea in psychology—creative problem solving centers on discovering and using a new analogy, and *that* equals *recollection plus reflection*. Inventing a new analogy between A and B (say, a bat's flight at twilight and a cracked teacup) means *recalling* B when you think of A. (Rilke compares a bat's flight to the crack in a porcelain cup. A creative discovery might be an engineering accomplishment—or a poetic one.) Just as an e-mail might remind you of a quart of milk you need to buy, a bat's flight might remind you of a cracked teacup. To convert this unusual recollection into an actual analogy, you must notice, think about, and remember it—that is, *reflect*.

A branch on the ground reminds you of, *recalls*, a skeletal arm and hand. On reflection, maybe you can use the branch to pull tennis balls under the fence from your neighbor's lawn back home to yours. A new thought. Not earthshaking, but a new thought.

When presented with a request for information (how far is it to Hartford?), we *recall* the corresponding information, or memories from which we can extract the information. The same mental operation that answers routine questions underlies the invention of new analogies—if I use it the right way.

The question then becomes: what made the recollection—*bat suggests teacup*—come to mind in the first place? After all, few of us ever invent such analogies. Creativity is rare, and it is probably the mental achievement we value highest. If new analogies drive creativity, what force drives the invention of new analogies?

The Origin of Analogies

We have traded the creativity problem for another that is better defined but just as hard. How do we invent new analogies? The philosopher Jerry Fodor wrote, in 1983: "It is striking that, while everybody thinks analogical reasoning is an important ingredient in all sorts of cognitive achievements we prize, nobody knows anything about how it works."[5] Not even, he adds, in an "in the glass darkly sort of way." No big developments, unfortunately, have changed the picture since then.

Of course, there is more to creativity than inventing analogies. Willingness to push your ideas to the outer limits and beyond is important. Ignoring limits and rules that are merely conventional is important. A feel for the elegance and economy of nature is basic. Curiosity is important; your capacity to be surprised is all-important. But nothing is more basic than the discovery of new analogies.

The problem is this: When I ask you, "What color is a sparrow?" and you recall the image of a sparrow and answer, "Brown and white," it's obvious how you recalled an appropriate memory. The question and the recollected image overlap. They both include the name "sparrow." (There might be other overlaps too.) But a bat at twilight and a cracked teacup *don't* overlap. There is a certain abstract resemblance that Rilke makes us see, but it makes no sense to suppose that this information is stored along with the memory of the teacup. (Would the teacup memory include an entry that says, in effect, "abstract resemblance to bat at twilight?" There are hundreds or thousands of resemblances at this abstract level, and they can hardly all be stored along with the cup memory.)

How *do* we accomplish this feat of recollection?

Recollection Using Essence Summarizers

If I could summarize the *essence* of a memory or experience—of a person, place, scene—I could stamp each memory with its "essence summary"—a sort of bar code or digital ID matrix. If two memories were each labeled XU5Z—same essence summary in both cases—I would know that they had something in common, even if they seemed very different. The same would be true if their summaries were *almost* the same, or closely related.

The mind's most effective essence summarizer is emotion. Two objects, persons, or events that are wholly unlike on the surface might make me *feel* the same way—or basically the same. And that similar feeling suggests, in turn, that these two must have something in common.

Hot coals make us feel burned—no thinking needed. Happy parties make us feel happy—no thinking needed. And we are (or *can* be) fantastically sensitive, delicate emotion readers. Two people who say nearly the same thing to us, in nearly the same way, can make us feel very different. It's no surprise that this highly refined skill should be useful and not just decorative. At our best, we are superbly sensitive emotiometers (that's "ee-moh-shee-AH-meh-terz").

You and I might be equally sensitive, yet our readings might be completely different. That doesn't matter, as long as yours are self-consistent and mine are too.

Imagine that I am at a fancy wedding reception. It amounts to "unpleasant chaos." In other words, the essence of the scene is unpleasant chaos. How did I fix on this *particular* essence summary—or verdict, or diagnosis? I might have done it by systematic thinking. Far more likely, the event registered directly on my emotions. We are sensitive musical instruments and the world gets distinctive sounds out of us by pressing the keys and tensing its lips.

If someone challenged me after the fact, I might present a checklist of attributes that made the wedding seem like unpleasant chaos. (Noise level. Quality of music. People I know. Activity in the room. Atmosphere. Level of drinking. Crowdedness.) But we rarely *decide* how an event makes us feel; we just feel that way. The event presses our keys, registers directly on our feelings. Human emotions are *essence summarizers*. They take us directly from a real-world situation to a *particular* emotion that captures, for us, the essence of the situation.

Now, the wedding reception might remind me of all sorts of other things that are superficially unrelated but share its essential quality. These other things might overlap the wedding scene in *no obvious way at all*—except in the shared essence. Maybe I recall a scene from a movie about gladiators in ancient Rome, or a high school graduation, or construction on a narrow New Haven street. The essence of any of those scenes might be *unpleasant chaos*.

VISUAL AND AURAL ESSENCE SUMMARIES ARE POWERFUL TOO

Suppose we are introduced to someone and soon afterward mutter to our wives, "She looks just like Sheila Bernstein." We have compared this new person to Sheila. We have performed some kind of image-to-image comparison to reach our conclusion.[6] In making our comparison, the mental image we use for Sheila *might* be a nearly literal memory of her appearance on some occasion, but usually we have something less precise in mind.

Clearly, the mind *can* work with visual summaries. Such a summary captures various aspects of the original and reduces them to essentials, boils them down, summarizes—with emphasis on what's *visually* important. We say, "the baby looks like his father," or "this actress must be the one who played Portia in that *Julius Caesar* we saw five years ago." Normally, we base such statements on a similarity between the person before us (the Sheila look-alike, the baby, "this actress") and a mental summary, a *visual* summary, of the other person. That we can *spot* this sort of resemblance (we can't all, but many of us can) means that *images are a medium that can be used for summaries.*

The same sort of essence summarizing can be done with aural sensations as well: "sounds like Brahms." One piece of music reminds us of another, or reminds us of a musical template. We might have in mind a template for the sound of Brahms. It holds no *particular* Brahms composition. It's an essence summary, in musical rather than emotional form. We might lack all explicit, conscious access to this template. We might not be able to describe it in words, or convey it in music, even if we *could* call the template to mind. But we *can* deal, however we do it, with a musical essence, synopsis, *abstraction*—of one composition or a whole body of compositions—of Brahms. What's important is that we can reduce the essence of a piece of music, taken as a whole, to a musical summary. *Music is another medium that can be used for summaries.*

But emotion is the most powerful essence summarizer by far.

Emotion is a hugely powerful summary medium, because an emotion can summarize *anything*. This means not only that emotions are *versatile* summarizers; they can be used to summarize people or places or cities, novels, historical epochs, butterflies. Far more important, emotions can be used to discover the *shared essence* (should there be one) of *virtually any two things we choose to compare*, no matter how radically dissimilar they *seem*. We can discover the shared essence (should there be one) between an aircraft carrier and a bowl of Wheaties. Between a neutrino and a can of Coke Zero, or a graduate program and the decayed medieval fabric of Old Saint Paul's in the center of London before it burned down, or asparagus and a scramjet.

The Summary Power of Emotion

An emotion is inherently abstract. It is *about* nothing. It has a cause, but the happiness *caused* by a faculty meeting's being over might be just the same (for you) as the happiness caused by a beautiful fall day. The collection of separate kinds (shades, tones, flavors) of happiness is boundlessly large. Each is abstract; none is an "emotional image" of any *particular* event. Take a photo of a wedding party, and you get a photo of *that* party. Take an "emotional snapshot" of the same event and you get a summary like "unpleasant chaos" that suits many different occasions.

Emotions, like colors, can be muted or shockingly vivid. Their variations are endless. Sometimes we feel several different emotions concurrently: happiness that Susan is coming Friday, worry that my laptop is failing and will be a nuisance to replace. I'm happy *and* worried. But they all add up to *one* final color, taste, emotion.

The plain fact of continuous consciousness creates many more emotional responses than we notice or remember: small blips or blats or dips or sudden spikes (too quick to register) of the emotiometer.

Lady Macbeth, as her husband arrives home from war, is ready for anything. "I feel now," she says, "the future in the instant." She *feels*—has an emotional or even physical experience. The future— she and her husband as queen and king—makes her *feel* some way. (Thrilled, awed, cruelly proud, smug, recklessly powerful.) The whole vision fits into one level *instant* because that's what emotions do. They are abstract summarizers. They can distill universes into teaspoonfuls of powerful essence. One essence, one abstract distillate: she feels the future in the instant.

Beethoven makes *one* statement about the mood in which some of his deepest music must be performed. *Etwas lebhaft und mit der innigsten Empfindung*: somewhat lively, with the *most* introspective feeling. One emotion can have many ingredients. *Langsam und sehnsuchtsvoll*: slow and full of longing. *Geschwind, doch nicht zu sehr und mit Entschlossenheit*: quick, but not too much, and decisively. No matter how many ingredients go into the mix, what comes out is *one* emotion.

"I feel pretty," sings the heroine in Bernstein and Sondheim's *West Side Story*. What kind of emotion is *pretty*? But you can feel *anything*. Everything you experience, you feel.

Emotions as Memory Cues—
and the Evoking Power of Mood

Emotions or moods can be powerfully evocative memory cues. They work like any other cue: if an emotion you experience *now* overlaps an emotion stored in, associated with, a recollection stored in your memory, you can reel in the memory. Material shared between cue and recollection makes remembering happen. An emotion you merely *think about* but don't experience can also be a cue. "Think of times you were insanely angry."

If you can't recall any aspect of a forgotten memory, it will stay forgotten. But sometimes an emotion will serve as a memory cue when nothing else will work.

You are unlikely to stumble on any particular detail that matches exactly a recollected scene from long ago—the same face or voice, sound on the TV or radio, passing traffic, passersby with the same sort of clothes or accoutrements—today, in the different world you inhabit now. But emotions are *abstract* summaries, and the particular emotion you felt and remembered long ago might crop up again, even though every sensory detail is different.

The uncanny ability of *mood* to reel in past time is one of the central facts about mind. Mood *is* emotion. Such recollections are likely to occur closer to the bottom than to the top of the spectrum. So we are *not apt to be paying attention*. We are apt to miss them.

This is the puzzle or (loosely speaking) *paradox* of mind. We have already considered it in a different context. The most interesting mental events happen at low focus, *because* of low focus. But you cannot (or can just barely) remember them, just because they *do* happen at low focus.

How do these mood cues work in practice? There are few examples on record, so I will supply some.

Gazing across the choir in a medieval church, at the opposite wall sunk deep in shadow, with the building dead quiet on a late afternoon and the gazer's wife and children temporarily invisible—gone off somewhere into the long, dark shadows— the gazer finds himself recalling a tour of the Château de Chillon near Geneva with his parents when he was five. On a cold winter's day, the place seemed empty except for his family and the guide. In a small room whose far wall was deep in shadow, he looked around—and the guide had disappeared. He was uneasy. (He knew about the castle's famous dungeon and doubted whether he could find his own way out.) "Where is everyone?" he asked (in just those words), and the guide said "*I'm* here," and stepped out of the shadows.

A silent, ancient, shadow-sunk building where the dark hides (or consumes?) people *feels* a certain way, creates a certain mood.

Here's another example in which Europe figures again in the cue.

In a fairly bare, austere Frankfurt hotel room that he is sharing with his son, the thinker has just awakened in the early light; it's quiet, and no one else is awake. There's a distinctly *I'm-not-at-home* feeling. Then he is thinking about the first morning's awakening at a summer camp when he was a child, in a room with four steel bunk beds in the four corners, bare and bright and quiet; he was, apparently, the only one awake.

The mood is bright, bare early morning in strange surroundings, with other beds in the room, and not knowing what to expect. At summer camp, I had no idea what the first full day would be like, and I would sooner have been anyplace else. I didn't mind being in Frankfurt, but I didn't know what to expect from a meeting scheduled for the afternoon.

The first example happened in late afternoon, when the thinker was tired and down-spectrum. The second is unusual: it happened during the low-focus period that intervenes between awakening and wide-awake alertness. Both show a mood serving as memory cue to retrieve an unexpected (eerily specific) memory of the past, an otherwise forgotten moment of past time.

Here are two other examples in a different key.

In the late afternoon I am working on an early stage of this book, feeling discouraged; then I recall myself on a particular New York subway platform at Penn Station in my early twenties–postcollege, pre-graduate school, studying art and Talmud in Manhattan, bound for Brooklyn to meet my parents at the apartment of my grandparents in Brighton Beach. It's probably Hanukkah. Part of the recollection is the unusual center platform, with the uptown express on one side, the downtown on the other. (You walk *up*stairs to reach this platform–which is nonetheless well below street level).

The recollection centers on a happy occasion, but there is a strong element of lonely sadness in my life at this time. It clings somehow to this particular subway platform. There is almost certainly something more here—mood-related—but I don't know what.

Here's another example:

Late one evening I came upon the phrase "I was nodding off *without knowing it*" while reading an earlier draft of this book; then I was thinking about the author Karen Blixen (pen name Isak Dinesen)–specifically as she appears in her published letters, at home in Denmark. The phrase *"without knowing it,"* its appearance on the page in italics, has something to do with the recollection.

My disappointment with Karen Blixen's (later) letters, and all the rest of her post-African writing, is connected to the idea of "nodding off without knowing it"—suggesting an unnoticed, unacknowledged weakening of one's powers. And again, there is more to it than I can account for.

One final example:

"Buried and seemingly out of the way, such a memory is sure to haunt you, Freud found, in the form of obsessions." This phrase comes, yet again, from an earlier version of this book. Rereading, I found myself thinking about the stairs leading to the front hall at my grandparents' in Flatbush–and remembered something I hadn't in a long time: a painting I'd made as a young teenager showing, from below, a man crouching above a pit–a deliberately strange and disturbing image. As a child I gave occasional paintings to friends, but most went to my parents; I wanted to keep them close. They didn't want any part of this one, though. Understandably. My grandmother, however, with her love of art in every aspect, was happy to take the painting, frame it, and hang it in her front hall. Superficially, the mood link has to do with Freud: my grandfather had published on Freud, and arriving at the Flatbush apartment meant entering a warm, welcoming scholarly enclave (as opposed to the more science-centered enclave of my parents' home). In deeper terms, the sentence quoted above deals with *avoiding unpleasant truths*: my painting presented itself as an unpleasant truth that my grandmother disdained to avoid–for which I was grateful, especially given that this painting, as I knew at the time, was awkward and in many ways unsatisfying. But still, I wanted to see it on a wall, not slumped or stacked in a dusty pile of rejects.

One might have described the *connection* between the sentence I quote and the long-ago childhood scene as an *idea*. Not a mood or emotion, but a type of thought. The connecting thought: "our tendency to avoid disturbing truth, and people who are willing to put up with it all the same" (Freud, of course, specialized in disturbing truths).

Yet I believe that the connecting link was, in fact, a *mood*, by which I mean a mood that captures or includes this idea. I believe it was a mood because the resemblance between the recently written sentence and the forgotten old scene in Brooklyn hit me out of the blue—not like a patiently made discovery; like a recognition. Like "*Of course* it's Sheila!"

I was thoroughly puzzled as to why the analogy had come to mind. Why did this memory of a long-ago scene suddenly occur to me in connection with the statement about Freud? Clearly, the mind *recognizes* two emotions' being close or identical, just as it recognizes identical faces or voices. Recognition is a kind of *seeing*, largely unconscious. Emotion, in short, is not only a way to connect two deeply related though superficially *un*related memories. It's a way to *recognize* such a connection unconsciously, almost instantly. Memory procedures that work *fast* are crucial when we need to find one memory in a large pool.

Here are the two basic points:

1. A mood or emotion can be a cue that summons memories that include this same emotion.
2. Such a cue can connect two memories that, in most ways, seem far apart.

Analogies and Creative Inspiration

Many books and papers have discussed examples of new analogies and the restructuring of problems, especially in science. I will merely sketch the possibilities with a few examples.

Winston Churchill coinvented the tank during the First World War on analogy, evidently, with the battleship. H. H. Asquith, prime minister at the start of the war and a brilliant man himself, described Churchill as "a curious dash of schoolboy simplicity" together with "a zigzag streak of lightning in the brain." Churchill, a member of Asquith's government at the start of the First World War, faced the problem of how allied soldiers could advance over deep trenches against murderous German machine-gun fire. When he met Major Thomas Hetherington, who had been thinking about the same problem on similar lines, the result was the "Land Ship Committee," and an active project to build such a vessel.

Here we have an analogy between a real object (the battleship) and a mere abstract set of requirements. No one had any idea what a tank would be like. But the *feel* of those requirements evidently brought the feel of a warship to mind: smooth, impregnable cruising; reliable protective shield; aggressive weapon.

My own first publication in computing, having to do with a new solution to problems of deadlock in Internet-like computer networks—deadlock in such networks is like "gridlock" in a large city—centered on an analogy between the movement of data packets in a network and of commuters in Grand Central Terminal. Grand Central has two floors' worth of train gates. Even if both floors were jammed up, one could imagine a vertically moving loop flowing down one staircase from upper level to lower and then back up another staircase. This analogy led straight to the solution. There is an abstract visual similarity, but it's *very* abstract. Emotion coding seems like the main ingredient in making the connection.

Finally, here's another case in a different domain. Rilke makes an astonishing comparison in the eighth "Duino Elegy" between an infant bat's or bird's first flight and a crack in a teacup. Poetic imagery exists to make us *see*, make us look again at what we take for granted and usually don't bother to see, to look at something old in a new way. Rilke uses the analogy to draw a conclusion: "So the bat's track / fractures the porcelain of evening." Where did this strange and powerful analogy come from? Presumably from abstract visual images that resembled each other. Emotion is the most powerful and general of the sensory-emotional summarizers—but not the only one. Visual summaries are powerful too.[7]

Do Emotions Really Work This Way?

Sometimes one can actually *feel* the feeling that connects two parts of an analogy. In Jane Austen's *Persuasion*, Anne must recover from her shock at seeing someone unexpected in the distance, and get on with her life:

> When she had scolded back her senses, she found the others still waiting for the carriage.

What inspired the author to write "scolded back"? To settle down, to calm oneself *feels*, in this context, like "scolding back." One *scolds back* noisy parties that can't be physically pushed around. One scolds back a group of noisy chickens or children; one throws up a screen of admonitions to settle them back where they belong.

In *La porte étroite* [*The Narrow Gate*], André Gide's narrator (who is in love) tells us, "Every day I was awakened by my joy." A simple, unpretentious yet nearly perfect sentence: *Chaque matin j'étais éveillé par ma joie.* No adjectives, no qualifications. I suspect many readers, over the years, will have said to themselves: *Yes, I have had that experience.* I have felt that way.

The sentence reminds readers of similar occasions in their own lives. How? Some of Gide's readers will use the feeling created by his sentence to go straight to a memory that evokes that same feeling. In which case, an emotion has arced the gap between Gide and you. You feel *along with* Gide. One thinks of Heidegger's idea of putting ideas "*in relation*."

"The heart is the seat of a faculty, *sympathy*, that allows us to share at times the being of another," says Coetzee's Elizabeth Costello. Notice: not "share the life," but "share the *being*." In spectrum logic, "being" means, again, sensation and emotion unconfined, allowed to fill the pool of consciousness and bring forth the moods and memories that follow naturally.

A musical note can create resonance elsewhere, sympathetic vibration. A sung note can make a nearby violin string vibrate. Emotion, too, can "resonate," can jump the gap between two people. John Carey discusses resonance in John Donne: "In Donne's [essay], orgasm fills the body like a musical note setting up its lingering whispers. By likening his girls to pure metal bells Donne suggests the secluded stirrings of their physical life: the metallic image renders them more alive, not less."[8] Bells awaken sympathetic vibration. One body picks up, or tunes in, or reexperiences, or *shares* another's being.

People used to say "a penny for your thoughts." No one ever said "a penny for your feelings." Your thoughts are often inscrutable, but your feelings are usually obvious. Sometimes we can *feel* (not just discern) someone else's feelings. We feel directly that someone is angry, uncertain, exultant, as we feel a warm breeze.

In all these facts we see the power of emotion to pull things together—two silent people, an author and reader who have never met and never will; two halves of a new analogy; two separate worlds.

The psychologist C. E. Osgood makes an important observation. The discovery of new metaphors, he says, is driven by the fact that "such diverse sensory experiences as a *white* circle (rather than black), a *straight* line (rather than crooked), a *rising* melody (rather than falling), a *sweet* taste (rather than sour), a *caressing* touch (rather than an irritating scratch) . . . can share a common affective meaning."[9]

Coleridge goes right to the point in a letter to his poet friend Robert Southey:

> I hold, that association depends in a much greater degree on the recurrence of resembling states of Feeling, than on trains of Ideas. . . . I almost think, that Ideas *never* recall Ideas, as far as they are Ideas—any more than Leaves in a forest create each other's motion—The Breeze it is that runs through them; it is the Soul, the state of Feeling.[10]

Clinical Evidence

Emotion as a memory cue is easy to understand. But does it happen in real life? Yes. Evidence comes from a distinguished psychoanalyst who was also a neurophysiologist, Morton Reiser.

Reiser reports, on the basis of long psychoanalytic experience, that far-apart memories (concerning very different topics at different times) can be "emotional analogues." He means that two memories can evoke similar feelings. He begins with Freud's idea of "memory nodes"—gathering points for many different thoughts with something in common, or many aspects of one dream. Reiser tells us that shared emotion can bring a crowd of varied thoughts or dream fragments together. Many seemingly unrelated aspects of one dream might all evoke similar (or the *same*) emotion.

"Contemporary data," Reiser writes, in reference to psychoanalytic experience *and* neurobiology, "implicate emotion as the glue that binds memory elements to each other, enabling those that belong together to stay together."[11] *The glue that binds memory elements to each other*; look *inside* an episodic memory, Reiser tells us, and you will find its various elements held together by their association with one shared emotion.

I believe that two seemingly different, unrelated thoughts might be joined—proposed by the mind as a new analogy—insofar as they evoke the same emotion. Reiser's view is different. Yet he, too, speaks of the *power of emotions to join* separate loci of the mind—separate thoughts, ideas, memories. "Dream images and the memory traces they represent may be associatively linked by a capacity to evoke the same emotion."[12] Dreams often contain strange images that seem meaningless—except to the dreamer himself, who knows immediately that the tall, bald man carrying a dolphin under one arm is actually a college girlfriend. How does he know? Where does the strange image come from? Sometimes, says Reiser, the image and the thing it represents—the bald dolphin man and the college girl—evoke the same emotion.

Reiser speaks of "an entire network of stored memories that are related to each other by shared affective potential."[13] He concludes that "*sensory residues in the mind are organized by affect and arranged as nodal memory networks*."[14] His "sensory residues" are memories. Within his "nodal memory networks," clusters of memories that *share an emotion* are linked together.

In Sum

The summarizing power of emotions is poorly understood. Yet it is, potentially, enormously important. I am thinking about a bat at twilight; then, suddenly, about a cracked teacup. And I write a poem that will live for as long as poetry exists. (Not based on this one image alone! But it plays its part.) I am thinking about a gridlocked computer network; then, suddenly, about crowds in Grand Central Terminal. The result is a distinctly minor discovery. But lots of minor discoveries make science. How does it all work?

This is such a challenging and important question that researchers have gone all out in proposing solutions. One eminent thinker, for example, believes that only quantum mechanics and the superposition of wave states can solve the puzzle. Isn't it more sensible to guess that I invent a new analogy when two different-seeming things make me *feel* the same way?

I haven't explained why and how feelings work as they do; my goal is merely to explain how the building blocks of mind—such as *emotion*, such as *recollection*—combine to create such phenomena as *creativity*. What I have done is help suggest the extraordinary power and versatility of our emotion-reading talents. We are champion feelers.

Creativity and new analogies are such hard problems that I must underline, in closing, what's been suggested. Creativity emerges when I compare a bat's flight to a cracked cup, a computer network to Grand Central Terminal, a still-unconceived military vehicle to a battleship. Simple enough. But how do I get the right pairs together? How do I find just the right memory to pair with my bat, my jammed computer network, or anything else, given the stupendously huge set of memories that exist within the brain's billion-odd neurons? "Your partner is somewhere on this dance hall; just page through these five billion CVs and pick her out!"

This is the problem of creativity. And the proposed answer centers on the mind's skill at labeling any experience with an *exactly* appropriate emotion—"exactly" for my own purposes, of course. Instead of billions of prospective analogy partners, I can sweep directly to the few (if there are *any*) that have the same essence summary as the object or experience I'm starting with. The human memory is a champion at pulling, from our huge collections, *just* the memory that is associated with the last time I walked into this shop, or the last time I heard this song. If memories are labeled, we are superb at finding the ones with the *right* labels.

Of course, I might still find far more potential analogies than I can handle. (And I might have heard the song too often for my previous hearing to mean anything.) None of that matters. Sometimes, we know, creative thoughts do occur, and often they center on new analogies; and now we have a plausible method for understanding the construction of those new analogies—a method that involves no previously undiscovered techniques or brand-new mind maneuvers. It centers merely on one of our hugely powerful pieces of mental equipment: our sensitivity as emotiometers.

We might now make some tentative progress on other questions too, such as: What is consciousness for? What is the point; what is gained? Why did nature (or God) bother with it?

In a sense it is a ridiculous question. No human being is willing to give up consciousness; life is meaningless without it. Consciousness gives meaning to life. That's hardly an insignificant piece of work.

Meaning has value to human beings. We need it. We lust for it. We are bitterly unhappy when we are forced to live life without it. But consciousness is a strange case. Without consciousness, our lives are without meaning. Would living without consciousness make us unhappy then?

Of course not. It would make us nothing at all. Without consciousness, we are neither happy nor unhappy. We are nothing.

Philosophers (some, anyway) are worried. What *is* consciousness for? "Zombies" intensify their worry. Suppose you built a robot that was an exact, perfect replica of a human being in every way *except* that it was not conscious. Does its lack of consciousness prevent its acting the role of a human being in any detail? Does the lack of consciousness matter to its performance in any way *at all*?

Seemingly not. When it says "I'm terribly unhappy," it feels nothing; it never feels *anything*, but it can run through its performance perfectly. We kick it in the shins. It says "Ouch! You damned . . ." "Why did you just say *ouch*?" we ask politely. "Because I'm in *pain*, you dumb bastard!" it replies. It's programmed to act exactly like a human being in every way, all the time. Its lack of consciousness doesn't seem to impede its performance in any way. Why should it? How could it?

And if not being conscious wouldn't impede our performance, why *are* we conscious?

Maybe consciousness does do something for us after all; maybe the unconscious zombie would *not* be indistinguishable from a human being.

A sung note is physically different from the same note on paper, or in our minds. We can hear the note in our minds precisely, without singing it. But the sung note creates sound waves—riles the air; might create sympathetic vibrations: make a piano string tremble or a wine glass quiver.

In very roughly the same way, a felt emotion is different from one that is merely recorded in your brain. You might remember, right now, a childhood occasion when you were terribly embarrassed. But you can recall your embarrassment without *feeling* embarrassed. *Knowing* about the emotion is different from *experiencing* the emotion.

Emotions (I have suggested) are crucial to certain sorts of recollection, including a sort that is essential to creativity. And *feeling* an emotion is plainly different from merely *being aware* of one. What are the implications of this difference for the process of recollection? It could be that if we *experience* an emotion—feel it, are *conscious* of it—the result is a more effective memory cue than if we merely think about the emotion. The *felt* emotion might conceivably reach, discover, activate, summon more memories—reach farther and deeper—than the unfelt, merely thought-about emotion. Some analogue of *resonance* might work in favor of the felt emotion.

Imagine a soprano waiting at the threshold of a large dining room (belonging to some sort of fancy club or restaurant, let's say), looking the place over from a step or two higher than the main floor. If she holds up a large sign that says, "Think of a high A-flat," she'll create a very different impression than if she *sings* a high A-flat. (Loud.)

If felt emotions are indeed (sometimes, in some ways) better memory cues than emotions we merely know about—and "felt," of course, implies consciousness—then a zombie would *not* be indistinguishable from a human being. A human would recollect things differently. The human being would be better at remembering based on emotion as a cue—and would therefore be better at (among other things) creative thinking.

Whether we need consciousness or not, we *do* need emotion—at least a faithful simulation of emotion, as in our zombie. We need to think, and therefore we need to remember, as human beings do. Without emotion as a cue to recollection, your memory would be a mere database, a mere computer. With it, miracles happen. We discover analogies. We create.

Seven

Spectrum, Lower Third: Descent into Lost Time

Three big principles, interrelated, shape the mind as we float gently, gradually, on the great soft wings of daydream into the lower reaches of the spectrum. Dreams are theme circles that pull the past into the present. Daydreaming, fantasy—mind wandering on a large scale—make a transition zone; then we approach sleep-onset thought, the hallucination, and the dream.

We live our lives as if in a backyard with a high fence on one side that we can't see over. Naturally, we wonder about that fence and what's on the other side. But we can't ever find out, and eventually we shrug it off and barely see the fence any longer.

On the other side, just out of sight, is the past. We live side by side with our hidden pasts, and never (or almost never) suspect. Right there, on the other side, is your life on some afternoon when you were eight, or thirteen, or twenty. That's where dreams take place.

In normal life we can sometimes just barely hear (when the wind is right) the play of hidden voices from the other side.

Daydreaming and Fantasy

Daydreams can happen anytime, but they become important in the lower-middle zones of the spectrum.

Eric Klinger, daydream specialist: "Day-dreaming *keeps reminding us* of our current concerns. . . . The concerns it comes back to most are those *emotionally most important to us*."[1] Daydreaming *and* dreaming are first and foremost remembering. Remembering is a process strongly biased—other things being equal—in favor of our newest, freshest memories. Daydreaming is biased the same way.

Daydreaming happens in the spectrum's lower half. But don't people daydream in the morning? Of course; we oscillate over the spectrum more than once in a typical day. For some of us, especially children, up-spectrum thinking is *never* natural. Children spend more time than adults in the daydream-rich regions of the spectrum. The association of daydreaming with the lower spectrum—no matter the time of day—is obvious: daydreams presuppose relaxation. To say you are "alert to your environment" *and* "daydreaming" makes no sense.

Daydreaming at *any* time of day means we are down-spectrum, increasingly outside the conscious mind's control. Sometimes we actively decide to daydream. But often, daydreams choose us. Jerome Singer, daydreaming authority, believes that daydreams are usually involuntary, tending to happen when our surroundings grow quiet or boring. Even when we do choose to daydream, often with an explicit goal in mind ("it would be fun to mull *this* over"), daydreams are like ordinary dreams in being *narrative*. Daydreams tell stories.[2]

What kinds of stories? Daydreaming as a topic often suggests "just suppose" or "if only" daydreams. When Elizabeth Bennet visits the magnificent country home of her rejected suitor in Jane Austen's *Pride and Prejudice*, she daydreams that she might have been mistress of the place. These are wistful thoughts, a bit sad. The author leaves us to fill in the blanks ourselves; she is not for lazy readers. (Which might be why Henry James, who eventually decided that filling in every last blank was his authorial duty and right, never understood her.)

But a short time later, the rejected suitor *himself* unexpectedly and embarrassingly appears, chats deferentially, winningly; departs again. More daydreams! Now Miss Bennet's emotions are all turmoil, and she is wondering, urgently, what he really thinks of her. Her daydream continues until "the remarks of her companions on her absence of mind roused her." This second daydream would have been pressing, perhaps anxious. Not pleasant.

Daydreams can be unhappy, and they are sometimes disliked by the serious-minded on principle. Anna Karenina's husband "regarded this mental activity as pernicious, dangerous daydreaming" (Tolstoy, *Anna Karenina*, Bartlett translation). Teachers have never been crazy about daydreaming students.

Daydreams resemble low-focus hallucination more than up-spectrum reasoning in being enveloping and engrossing. (Miss Bennet has to be "roused.") But sometimes the conscious mind plans and steers daydreams explicitly, which makes them unlike dreams—where consciousness plays an important but largely hidden, implicit part. Daydreams, in short, have some down-spectrum and some up-spectrum characteristics—just as they ought to. They are lower-mid-spectrum creatures.

Daydreams are *not* hallucinations. But they can come close. "When you approached her," writes André Gide, as I've noted, of a daydreamer, "her eyes would not turn from their reverie to look at you" (*La porte étroite* [*The Narrow Gate*]). We all know such people and moods. "Look how our partner's rapt," says Banquo of daydreaming Macbeth. "Worthy Macbeth, we stay upon your leisure."

Macbeth's tendency to raptness is crucial to the man and his personality, and therefore to Shakespeare's story. Macbeth is called "rapt" three times in early scenes. His psyche is the strangest Shakespeare ever invented—but it hangs together perfectly. He is a low-spectrum character, a seer and a visionary; but he also has high courage, a weak character, and a grasping wife he adores.

We know that thinking visually grows more important down-spectrum. Far down-spectrum, just outside sleep, we encounter the hallucination line. Now suppose we lop off the entire upper half of the spectrum and substitute the lower half for the whole. Visual thinking is more important in this truncated spectrum than in the whole, untruncated spectrum. We reach the hallucination line sooner in the truncated spectrum, relatively speaking, than in the untruncated version.

Macbeth has a powerful visual-thinking bias. To think something is to see it. And the pictures he imagines often boil over into a scalding mist of hallucination. Of course, Macbeth is a master of words too. The fantastic power of his visual *and* linguistic imagination suggests one other man only. A certain playwright.

A daydream is easily interrupted. A daydreamer is easily roused, but daydreams can be engrossing. Naturally, sexual fantasies can be even more so.

Chateaubriand was ardent and deeply imaginative, an impoverished nobleman, a romantic, swashbuckling monarchist during the French Revolution, a daring fighter, a daring author. An original all through. As a teenager in late-eighteenth-century, pre–Revolutionary France, growing up in the family's ancient, beat-up château in the Brittany backwaters, he faced a stiff challenge. He desperately wanted, he *needed*, to fantasize about girls, but he found it hard because—growing up in strict circles, in strict times—he had never met any (only children and his sisters excepted). But he rose to the challenge. He *invented* a girl, working only with the meager data at his disposal. Then he worshipped her devoutly every day. His imagination was gigantic (Chateaubriand, *Mémoires d'outre-tombe* [*Memories from Beyond the Grave*]).

In Chateaubriand's adolescence (he tells us), he was impatient to get to bed, where he would unleash his majestic fantasies. "All the powers of my soul were exalted to a state of delirium. . . . The world was delivered into the power of my *amours*."

Daydreams are visual, narrative, engrossing—in other words, down-spectrum. Yet we control them deliberately (at least sometimes), we are recognizably our own *selves* in them, we remember them—not perfectly, but far better than we remember dreams. We have descended a long way down-spectrum, but there is farther to go.

Dream Thought

When we dream, memory has the floor. Conscious mind still has the power to reject highly upsetting memories. But most memories are admitted when we are near or at the bottom. The price we pay is the occasional nightmare, and frequent vague upset or uneasiness. Nature balances the books by arranging for us to forget nearly everything. What we don't remember never happened to us. Except, forgotten dreams can have a subtle influence over waking life.

Happy dreams (especially sexual ones) will sometimes burn bright for many hours. On the eve of Saint Agnes, when maidens were said to dream of their lovers, Keats's Madeline looks forward as she climbs into bed to "all the bliss to be before tomorrow morn" ("The Eve of St. Agnes"). (To all the blissful *dreams*, that is.)

James Joyce's alter ego awakes after a good night:

> An enchantment of the heart! The night had been enchanted.
> In a dream or vision he had known the ecstasy of seraphic life.
> Was it an instant of enchantment only or long hours and years
> and ages? (*Portrait of the Artist as a Young Man*)

We don't have to remember anything *about* the dream except that it was erotic or good in some other way. Joyous dreams cast their glow whether we remember the details or not.

"Enchanting" dreams don't crop up often. But ordinary dreams can affect daily life too, even if we have forgotten them almost entirely. By chance, we do or say something related to an almost-forgotten dream—and we feel a faint answer; a mysterious pale answering glow beneath the sea surface of memory. Elements of the dream tremble or tingle when we land on a thought (like a lucky square in a board game) that matches some part of the dream.

More common than an accidental memory is a moment of awareness too vague to be traced. We pause for a moment, for reasons we can't explain (we rarely try; the pause itself usually goes unnoticed). We merely stop for an instant; that's all. All we feel is the slightest quiver, the slightest sense of something somewhere answering back. But these little unnoticed pauses, graceful quarter rests in the rhythm of life, help us live.

If we choose to confront our dreams or sleep-onset memories, we will learn what is on our minds. Sleep-onset thoughts are especially valuable in this respect (though harder even than dreams to monitor), because they are apt to be less distorted. They can take us way back into the past, into early childhood.

Dreams are governed by several principles.

1. Memories of recent events come first.
2. Memories that appear in a dream record events of the outer *and* inner fields of consciousness—memories of external reality and memories of our own thoughts. But dreams speak in pictures. Words appear in dreams, but we distrust them. So words take a secondary role. When we remember an idea in a dream, it is translated into a picture, and *then* we remember it.
3. Dreams are theme circles. In fact they are often multiple cir-

cles, with some elements of a sequence belonging to one theme, others to a different theme.

A theme can be anything, but it is almost always an emotion-steeped image. The emotion is most likely to be a *blocked* emotion— a smoldering, unresolved emotion—because those are the strongest around, the stiffest drinks in all of memory.

Remembering Creates the Bizarreness of Dreams

The same construction principle that leads to illogical dream sequences leads to illogical dream scenes and dream images.

The principle is simple. If a dream includes a strange or illogical sequence (you're at the beach as a child, then looking for a space in a parking garage near your office), that same illogical sequence yields an illogical *image* if the two scenes appear concurrently, superimposed, or jumbled together, instead of one after another. Instead of the sequence "beach, parking garage," you would see an image that is part beach, part garage.

We know that a memory cue can yield many recollections. And we know that *conflation* of separate recollections allows us to create abstractions. It doesn't seem unreasonable to suppose that separate recollections might also be conflated during dreams.

Memory presents us with an armful of close fits to a search cue, in no special order—except for a consistent bias in favor of emotion. Suppose my search cue is an image that I have just seen: a macaw stretching his left wing and right foot simultaneously, as he likes to do. Lots of memories overlap the cue. Some will be visual; some will overlap an emotional summary of the little scene.

At low spectrum, your memory is pumped up full of energy—

as is your conscious mind at *high* spectrum. Your memory is eager to go to work: to fetch recollections based on cues, to gather the best or closest-matching recollections and push them straight into consciousness—the whole process without waiting for *any* requests from the conscious mind. By the same token, at high spectrum our conscious minds jump eagerly to work, solving problems and making plans, without necessarily waiting for an explicit problem or request.

So our memory cue is a bird image. Memory responds by marking or putting forward recollections that match the cue. The closest, best matches are pushed forward into consciousness.

Sometimes memories are pushed in a bunch into conscious mind, instead of going one by one in a neat, orderly fashion. If memory grabs recollections A, B, and C all at once and shoves them en masse into consciousness, the result is an overlay. Memory knows how to do overlays, as it does in creating abstractions or templates. But in this case, the overlay's separate elements have nothing to do with one another *except* that they all answered the same call. Assuming that these are all images, the resulting overlay is visual nonsense, which your conscious mind does its best to understand.

This point is important; the bizarre images and scenery of dreams are one of their most striking characteristics. Freud has an elaborate explanation in terms of the "dream work"; many modern dream researchers attribute the strangeness to pure chance. It *is* chance, but not pure. Not at all pure. The recollections that are awkwardly overlaid into strange images, and intertwined into strange sequences, *have all responded to the same cue.* It's a subtle relationship. But it is definitely a relationship.

Dream Themes

There are endless possible themes, but certain themes are ubiquitous as foundation ingredients in other emotions. Of these, the most important is a special kind of homesickness, for a home that no longer exists and never will again. ("I was homesick, had been homesick for months. But home was hardly a place I could return to. Home was something in my head. It was something I had lost" [V. S. Naipaul, *Bend in the River*]. "Everyone alive mourned the loss of his home-world." [Saul Bellow, *Humboldt's Gift*].)

"Lost-home-sickness" (homesickness for a lost home) is a sign of good luck, in a way; those who experience it think of their past lives with love or at least fondness—or at very least, nostalgia. But we find lost-home-sickness even among those whose childhoods were rough. It is a powerful and nearly universal impulse—one that underlies not only our own recollections but our collective memories of the good old days, and golden ages past.

Dreams fulfill the deepest of all human wishes: to go home. We can't. Yet we can and do, on the inside, every day.

The Dreamworld: Why We Forget It

"Overconsciousness" means that we lack the resources to create solid memories; we lack a proper sense of self and have no assets to spend on handling and hardening brand-new memories so that they last.

But there are other reasons why we forget our dreams. For one: our companion time thread breaks when a dream starts. Since late infancy or earliest childhood, we have been aware of the steady progress of time. We all have a mental clock. Think of it as a tape measure we drag constantly behind us. It rules off the past into minutes, hours, months. It doesn't reach terribly far back. Usually it starts to blur out and disappear beyond a few months into the past. But it's often useful.

One reason it's hard to remember dreams is this "tape measure." When we look into the past, we follow the route of the tape. We can easily say what we were doing (roughly speaking) ten minutes ago, or half an hour or three hours or two days ago. But the tape measure doesn't cut through dreams; it steers around the outside like the beltway that avoids downtown traffic. If it's eight in the morning and we ask what you were doing three hours ago, you will probably say "sleeping." You will almost certainly *not* say, "battling a giant goldfish in a dream." We have no consistent, continuous measure of time that reaches *into* our dreams. Each dream inhabits its own separate world, with its own separate clock.

Each dream's timeline is separate from the one that runs alongside our lives. Each dream has a timeline that runs *parallel* to that master timeline, in the sense that (at least for the brief lifetime of the dream) the two timelines, the dream's and the other, never meet.

There's one last simple reason why dreams are hard to remember.

Each dream defines its own world of experience; we aren't normally reminded of a dream by looking out the same window we were looking through during a dream, or speaking to the same person, or sitting in the same chair—because we are usually transported into strange dreamworlds that have their *own* scenery and furnishings and, in a sense, people; dream people don't quite align with real ones.

But it's not merely that we are located in a different house or a different street or lawn or sidewalk from the ones we know in reality. Imagine that you have just barely awakened from a dream—*so* just barely that you can still look right back *into* it, as if you had just stepped out of a tunnel (say, the enclosed walkway that connects your plane to the airport terminal) and paused to look back in.

What do you see, looking back *into* the dream from reality? You see an entire world that is just slightly, but noticeably, *misaligned with reality*. As if the entire dreamworld were tilted slightly on its axis—say, by fifteen degrees. Enough so that *nothing lines up*. Even if your dream were set in your own bedroom, as you lay in your own bed—as some dreams are—the texture of every surface, the feel of every object is *different* in a dream. After all, a reality made of perceptions has been replaced by one made of recollections. There *must* be change!

You will see this yourself if you look back soon enough, before the whole thing has "melted into air—into thin air." Recall that Shakespeare put this famous line in the *Tempest* to describe the dissolution of dreams.

Transitioning to Dreams via Free Association

You know that you are no longer merely drowsy, that you are sliding down the slipway toward sleep, when you notice a thought hanging around consciousness (perhaps modestly, in back) that *you* didn't put there.

Often we become aware of thoughts at the edges of our dreams, seemingly ripe to be missed, just as we notice things at the edges of waking consciousness. (A certain recollected garden pavilion on his father's estate in vanished imperial Russia "hangs around, so to speak, with the unobtrusiveness of an artist's signature," writes Nabokov about his adult dreams. "I find it clinging to a corner of the dream canvas, or cunningly worked into some ornamental part of the picture" [*Speak, Memory*].) That "I didn't put it there" observation marks the start of free flow or free association, and your steady descent into sleep and dreams.

Let's return to your lost keys. After some daydreaming in midafternoon, more hours pass and now you are home at the end of a long, exhausting day. It might be about nine o'clock. You find that your keys are gone. You're annoyed, angry. Maybe you check your pockets mechanically and cast around the room—attempting to solve the problem with a sheer minimum of thought, or none. Maybe you yell at the kids or sulk, making no attempt at a rational solution.

Now let's move the whole scene later: 11:30 and you're sleepy, about to go to bed. Or maybe you already have. You notice that your keys are gone; what to do? First, maybe, sigh and take a long look at the ceiling. Next moment, you're thinking about a day last spring when your wife lost her keys. (You didn't *decide* to think about that day; the thought just happened.) Then, a childhood afternoon when your father accidentally dropped a sheath of papers into the water as the family walked off a ferry. You make the effort to haul back on the heavy net of thought. Once again, you think about your keys. Soon after, you find yourself (again) far away.

Eventually, thought starts to flow. You start to free-associate—automatically, unconsciously—although you are conscious of each stop along the way. Memory is taking over, putting thoughts into consciousness.

Below the daydreaming regions, we enter the realm of free flow, which leads to sleep. This free flow is just free association, but I use the term "free flow" to make a distinction. A friend or psychoanalyst might ask you to "free-associate," and you might oblige. Given a starting point, you say the first thing that comes to mind. Then, from this *new* spot, you leap again. Then again. You have made your mind blank, as far as you can. You try just to leap, not think.

But this *deliberate* free association, with the conscious mind intervening to help each step forward, is *not* free association. It is, at any rate, different from the flow that occurs as you approach sleep—although it can *cause*, can *turn into*, real free flow, which happens entirely by itself, at no one's request, and without any helpful intervention by the conscious mind. *Deliberate* free association is, finally, a contradiction in terms.

Later those thoughts will grow in vividness and become whole scenes—at first static and brief. That will be "sleep-onset thought."

Finally, they will become more elaborate, the breaks between them will close, and the conscious mind will need to struggle even harder to get *meaning* from them. It will struggle to make them into stories. That will be dreaming.

Let's look at an example of free flow. A young boy narrates, describing his experience as he sits and thinks during dinner:

> First came the vacation and then the next term and then vacation again and then another term and then vacation again and then again another term and then again the vacation. It was like a train going in and out of tunnels and that was like the noise of the boys eating in the refectory when you opened and closed the flaps of the ears. (Joyce, *Portrait of the Artist as a Young Man*)

Three ideas form this flow: vacations and term time; trains and tunnels; sounds of the crowded dining hall. The first suggests the second, which suggests the third. The thought sequence as a whole *goes nowhere*. It's a simple theme circle; it curves back on itself. Each of these three ideas is a variation on a theme: *alternation*. We could think of them as a loose three-point circle with "alternation" in the center.

Free flow is like ordinary daydreaming—only more so: more sustained, less under control. We are less able to rouse ourselves. We are approaching sleep. The "key" to the state of sleep itself, Foulkes notes, is "relinquishing voluntary self-control of ideation."[3]

Relinquishing it to *what*? To the unconscious mind. *Memory*, the unconscious mind, merely pushes the latest retrieved recollection into consciousness. And there, emerging into consciousness, is our next topic of conscious thought. When an entire thought or recollected scene is taken as a memory cue, any element can be used to fish out more recollections—or the *whole memory* can be used as a next cue, via a powerful summarizer like images or emotion.

In these periods, your thoughts often flow. But sometimes, separate thoughts are framed in silence—when you think no thought at all. The mind is empty. Other times you have the impression that, whenever you scoop a hole in the sands of time, your thoughts fill it like seawater.

Blocked Emotion: The Main Creator of Theme Circles

Emotions are blocked when, as Freud explains, we refuse to let them express themselves—refuse to let the dissonance resolve. We turn our backs on them. Sometimes we ban them not only from waking thought but from *all* thought. A blocked emotion that cannot speak even in dreams resorts to other methods of expressing itself.

But it's likely that blocked emotions are allowed into low-spectrum thoughts (not just in dreams, but in sleep-onset thoughts and sometimes even in free flow) far more often than they are turned away at the door. After all, blocked emotions dominate our dreams. Neurotic symptoms are indeed common—but most of us are largely free of them. Common sense suggests that, by and large, blocked emotions *are* allowed to speak, but only in the relaxed conditions of the lowest spectrum.

Where do all these blocked emotions come from? In a sense, every emotion aroused by the past is blocked. Nothing stops us from expressing it *now*. But the scene itself is over, gone forever. In most cases we are feeling emotions we did *not* feel or express (not in their current form) in the past. We missed that chance. They are forever incomplete: fuel unburned; potential energy unspent. They dominate the lower spectrum.

Why should these emotions be important? We are attracted to strong emotion, and we seek out patterns—always. We seek *mean-*

ing, the opposite (in a sense) of randomness or chaos. Blocked emotions are the greatest *organizers* of thought in the lower spectrum. In those regions, memory is liberated to shove recollections into consciousness; thoughts flow free. Strong emotions, especially *blocked* emotions, are sheepdogs that organize the woolly, baaing, free-roaming thoughts and recollections of the lower spectrum.

Free flow seeks a theme as it moves along. Again, we *need* themes, we need *meaning*. We see the process operating in our dreams: we notice, sometimes, that the plot seems to be improvised by conscious mind as we go along, to fall in with some theme. *A dream seeks a theme* as liquid seeks the lowest level. Thought allowed to flow free on its own *always* seeks a theme. Humans are meaning-seeking creatures. After we have found food and drink and shelter, sex and companionship, we seek *meaning*.

How does free flow hit on a theme? The magnet that started the flow running is almost certainly an emotion. Strong emotions in the area, often associated with images, are good candidates for a theme.

Memory pulls out recollections that match the theme and pushes them into consciousness. There might be several themes, working sometimes individually and sometimes together. But we need to see actual sleep-onset sequences and dreams to see these principles operating.

Sleep-Onset Thought: Hallucinatory, Theme-Circling Thought

Toward the bottom of the spectrum, we enter the zone of "sleep-onset thinking," in which some recollections are replaced by hallucinations. The elements of the free flow grow more vivid and complete. Each becomes a separate scene, expressed as an image. Often these images are hallucinations.

Sleep-onset thinking has been found to consist of several distinct phases. With each successive stage,

> there was a steady decline in control over the course of mental activity and an awareness of the immediate environment and a steady rise in the frequency of hallucinatory experience.[4]

By some measures of content and quality, sleep-onset thought is closer to waking fantasy than to the dream sleep that immediately follows it. On the whole, researchers can distinguish (with some errors) sleep-onset-thought reports from dream reports. But other tests confirm the "essential similarity" of sleep-onset-thought and dream reports. Once again, the spectrum is a plain fact. Sleep-onset thinking comes *between* waking fantasy on one side and dreaming on the other, showing resemblances to each.

Hallucinations arise first in sleep-onset thought. This is where we cross the hallucination line. Where do hallucinations come from? Of your memories, some are "episodic," remembered scenes or experiences, as opposed to remembered skills or procedures (how to read, ride a bicycle, buy a drink).

Each episodic memory is, potentially, an alternate reality.

Sometimes remembering "the beach last summer" means *reentering* the experience: seeing the water, feeling the sand and sun, hearing the shouts and crashing surf; falling through the thin, brittle crust of memory (like the surface of newly dry lava) into the red-hot stuff itself.

EXAMPLES: SLEEP-ONSET THOUGHT

Let's consider a typical sleep-onset example, experienced as a series of separate hallucinations. I have already discussed this example, but there's more to say.

(1) Our macaw stretching his left wing. (2) Ping-Pong. (3) The iridescent colors on a pigeon's gray neck. (4) Rabbi S. in a car; he is driving. (5) Rain on Yale campus. (6) The smell of spring rain on campus and on the streets of Flatbush, Brooklyn. (7) C.'s long, dark, silky, fragrant hair; Palestrina, Rilke. (8) A feeling of turbulence in which many memories are dissolved.

The real theme here is a blocked emotion. In fact, *the* blocked emotion: lost-home-sickness. Here I am missing my college years, and my two pairs of grandparents and their homes in Brooklyn.

The sequence has a characteristic *look* too, a visual theme: a pearl-clouded opalescence. The mist-gray pigeon with its iridescent neck feathers fits perfectly.

My mother's parents lived in Flatbush—in the 1960s, it was an ordinary middle-class neighborhood. Their area was full of sturdy, four-square, two-family houses built between 1900 and 1930: tiny front lawns carefully tended, small backyards, one lot separated from the next by the width of a driveway (two tracks of concrete) leading to the garage in back. Houses of unhealthy-looking yellowish stucco and purplish-red urban brick, smooth and cold as metal. Too many dogs. Flatbush dogs were all small, with the air of hardened criminals.

But the lush, towering rows of huge sycamores between sidewalk and street redeemed everything. My grandparents were renters and lived up in the treetops on the second floor, eye to eye with the canopies (landlords were on the first floor). The living room had a line of windows along the front, and the leaves surrounded you. In April and May, the festive yellow green of the sycamores *was* the spring, and the trees drizzled their winged seedpods all over everything.

On important Jewish holidays the streets were full of middle-class families dressed up, with their children and pretty daughters, and sometimes their grandchildren too, all headed to various synagogues. Right on Avenue I to the liberal one, left to several that were orthodox—everyone greeting everyone else.

During my freshman spring in New Haven, I watched from a high-overhead dorm window as the hard-beaten yard of Old Campus was reseeded and the garden edgings of the buildings and yards exploded into daffodils. I felt, as always, part of and not part of Yale simultaneously.

In the lower spectrum, all memory trails lead to the past. As we fan out through memory, casually beating the bushes, we quickly come upon some strong, distinct emotion that becomes the center of a theme circle. Here there are two related themes, joined by spring rain. The macaw memory, which must have come from only a few minutes earlier, leads to a sequence that is quickly captured by one emotion theme and then another, long-ago Flatbush and long-ago Yale. The remembered images swing round the two themes like spaceships captured by a large object's gravity (two large objects in this case) and slung into orbit.

The thoughts in this sequence circle the theme like a maypole. Theme-circling thought restates the theme repeatedly in a series of variations. Ping-Pong, rain on campus, and lovely C. are views of Yale. Rain in Flatbush, and probably the pigeon, are views of Brooklyn. Ping-Pong in element 2 was something we did as undergraduates, and also probably refers to my sense of bouncing back and forth between home and away—two states of mind.

Sleep onset and dreams are made of imagery. Language can play a part, but in dreaming and sleep onset we have no command over language, as we have none (or little) over our own memories. We encounter phrases seemingly overheard, not invented on purpose. We state our themes in imagery. Imagery seems as obedient as language is fractious. You will notice, if you watch carefully in dreaming or sleep onset, thoughts in the conscious mind forming and being turned into images as they emerge from the chrysalis.

Images can be richer than words, but they are often ambiguous. An image plus an emotion, not the image alone, is the low-spectrum mind's proper replacement for language. Not the *equivalent* of language in any sense, but a different way to record and remember ideas.

Here's another sleep-onset example, in which the unconscious shows off its cleverness. I had recently seen Bizet's opera *Carmen* on DVD.

[1] Carmen's aria at the start of act 2. [2] Shampoos. [3] Sixth Avenue woman "lost her nerve"; falls from high building. [4] Vacuum cleaners. [5] Beehive–laundry clang [washing machine bell or buzzer]. [6] Korean children–computers–"they don't understand their language." [7] Raffael [spelled just like this]; Vittoria Colonna. [8] "I'm not an unstrategic doer"–companionways on a navy ship. [The word "companionways" means stairways leading below from the main deck, but I pictured them connecting many levels of deck and superstructure, running outside.]

A series of separate, brief hallucinations. Sometimes detail has been lost in the transcription (which is based, like the others, on brief, half-asleep, spoken descriptions). The quoted phrases occurred as out-loud, overheard language in the hallucination. But I pictured the word "Raffael" as written, with two *f*'s instead of *ph* as in English, and without the final *o* used in Italian.

The sequence is incoherent in itself—although as a whole it points clearly to a theme at the center of the circle. Note that the incoherence of individual images is made of the same stuff—reflects the exact *same* phenomenon—as the incoherence of the sequence itself. This is important to our understanding of the bizarreness of dreams. In a dream, a spurious narrative would be stretched over the whole framework, and it would be harder to see the underlying architecture.

Obviously, vacuum cleaners have nothing (or nothing much) to do with beehives and the buzz or ding of washing machines. The sequence is just an accident. Memory has shoved vacuum cleaners into consciousness followed by an image involving beehives. The beehive image is—by the same token—an obvious result of two separate memories (beehives, washing machines) being shoved into consciousness simultaneously instead of separately, in sequence.

In a dream, the plot would be worked up in such a way that the washing machine and the beehive were somehow related. But in sleep-onset thought, the conscious mind has yet to enter its dream-spinning state. Storytelling comes into full bloom only at the bottom of the spectrum.

The bizarreness of dream imagery isn't merely a matter of strange combinations or superpositions. We sometimes see things in dreams that don't exist anywhere else. For example, a wraparound sales counter of the sort used in department stores for displaying watches, jewelry, and so forth (customers around the outside, sales-clerks inside), except the counter is replaced by a narrow, trough-shaped, chest-high swimming pool—a rectangle whose edges are made of swimming channels—in which customers are invited to take a brief, refreshing dip. Or a roller-coaster car whose left and right halves sometimes run together and sometimes veer apart. The car itself looks "like a Gehry building or maybe a stadium from the Chinese Olympics," my notes say.

My guess is that these strange items are simple attempts to make sense of odd combinations of images, recollected concurrently by accident. A sales counter *and* a narrow canal, or trough of some sort, or swimming pool. We *experience* these images; we always make the best of experience. When we round the bend as we drive north on I-95, we don't say, "That city *can't* be there to the north of the highway; this must be some sort of brain misfire." We figure out the scene as best we can and do the same when we dream.

Let's return now to the sleep-onset sequence. It is a classic theme circle around a blocked emotion. I had shouldered the emotion out of waking thought because I resented it's even arising. It was a warning against something I was in no danger of doing—had *never* done, will never do. There was something bookish and conventional about the warning; it was connected to things I had been reading and writing, not doing. In any case, the blocked warning was this: married men do not get involved with pretty young women. Many thanks to my unconscious mind for this vote of confidence. I include it because it is such a good example of memory's plotting and planning.

One act of *Carmen* (although not act 2) takes place on a high mountain pass. In our film of the opera, it looks dangerous. One could easily fall. Danger of falling connects element 3 to *Carmen*. So does the phrase "lost her nerve" (notice "her"); Carmen *never* loses her nerve. Daring is one of her striking attributes. (Danger of falling connects Carmen and the companionways too.) Carmen and her lovers, Don José and then the bullfighter Escamillo, make a Latin-sounding pair, like the one named in element 7. "Shampoo" has to do with washing and thus laundry, as in element 5; it also has to do with a young woman I had been friends with, K.F. "Sixth Avenue" probably refers to an essay by E. B. White about the old Sixth Avenue el. Along the way you could see right into second-story windows, and I associate this scene—which suggests 1930s–New York paintings by Edward Hopper—with laundry hung out of windows to dry.

To start, the theme seems to be *danger*: danger of falling, for one. A beehive can be dangerous too. Likewise "not understanding the language." But *Carmen* is about a *specific* danger: of falling for an unprincipled, seductive woman (and of leaving the person you love, as Don José leaves his fiancée for Carmen).

There was nothing at all unprincipled about my friend K.F. But element 7 is the giveaway; it tells the whole story. K.F.'s last name is related, by a single memory hop, to one of the names mentioned in element 7.

We notice now that element 7 includes two glaring *mistakes*. Raphael (or Raffaello) is spelled wrong, any way you look at it. And Vittoria Colonna is famous as *Michelangelo's* friend, not Raphael's. Theme: "Danger: you are making an *obvious* mistake!" (Obvious to all those who are obsessed with Michelangelo and know a little about his life. Hardly a small group.)

This element of the sleep-onset sequence was a warning: "Don't even *think* about it." My friend's last name was related to one among these paired two-part names; and ordinarily, I spend much of my time painting—like Raphael (in a sense). Furthermore, literal-minded dreams and sleep-onset memories might easily misunderstand "companionway" as a walkway for companions. They sound dangerous, like "lovers' lanes."

There is more to this sequence—some of which I understand, some not. But the sense is clear. The theme is clear, and the theme *circle* is clear.

Dostoyevsky operated down-spectrum, in the realm of intense and saturated imagining. Dreams were important to him, and so were dream techniques. "It is a peculiarity of the narrative structure of *The Brothers Karamazov* that a given motif will appear in a number of variations, and a given moral or psychological theme will be represented by a number of different motifs."[5] Thus a theme gives rise to a family of motifs, and each of those appears in a series of variations. For my purposes, these are theme circles: motifs circled by themes, each circled in turn by variations.

One final example, with another pointed theme:

(1) Birthday in Briarcliff (aged ten): my Grandpa Sam buys me two miniature toy rifles in town (Ossining). (2) Backyard in Briarcliff, summer; we're going into White Plains for dinner. (3) The large Briarcliff pool late afternoon; getting out of the car, wearing plastic pinned-on tags, walking over to the outdoor shower at the pool's gate. (4) Pet shop in Ossining at the foot of the main street; yellow plastic sheeting in a display window nearby; also nearby, the barber who had a big stack of *Playboy*s but wouldn't let us small boys read them, though we often tried. (5) Croton Station; to the city (Brooklyn) by train with my Grandma Bea: she shows me how to slip the ticket into the clip for the conductor to punch.

We lived in Briarcliff—a village in Westchester not far from New York City—until I was twelve. The towns of Ossining and Croton were nearby. Element 4 is about frustration: I wanted a pet, but my parents didn't. I wanted badly to look at those *Playboy*s, but the surly barber, who distinctly disliked children (or at any rate, me) wouldn't allow it.

Notice element 2: one reenters some past moment entirely, and the future seems just what it seemed then. We were going out for dinner that evening—and that long, long ago dinner (of which I have no memory, as far as I know) still lay in the future.

The theme is a blocked emotion. I took for granted, paid too little attention to, my father's parents as opposed to my mother's. My father's parents never went to college, and they both spoke with Yiddish accents. My grandfather spoke English well enough, but he had learned it as an adult and was never comfortable speaking it. My Yiddish was near zero. My grandfather worked hard all his life, as a waiter, cook, manager, and everything else but *owner* in kosher delicatessens and vegetarian restaurants in Brooklyn. My father's parents had less money than my other grandparents (who didn't have much either), but they indulged me. They liked being with me, and I with them. But I didn't do it right. I didn't do it well enough.

I *was* close to my Grandpa Sam and Grandma Bea, but even closer to my mother's parents. My Grandpa Sam was kindly, gentle, taciturn, strong-muscled; he had come to America after serving in the Austro-Hungarian army during the First World War. I never talked to him as much as I wanted to, or ought to have. My Grandma Bea was comfortable with English, but she, too, had a pronounced Yiddish accent. (My other two grandparents spoke fluent, exemplary English.)

On my birthday once, we walked toward evening down main street in the hilly, riverside town of Ossining, and my Grandpa Sam bought me just what I wanted, what I pointed out to him—some miniature toy rifles, copies of Revolutionary War weapons. As always, he said little, but he was quietly, warmly pleased that *I* was so pleased. At the Croton terminal once, I got on a train with my Grandma Bea, bound for their apartment in Brighton Beach near Coney Island. She showed me what to do with the ticket, explained the transfer to the subway, and so forth.

The theme of this circle tells me: never forget how much they loved you, and how much they knew and taught you. This is not a *blocked* emotion; I would never refuse to allow it into consciousness. But it might as well be blocked. It appears constantly in my dreams and sleep-onset thoughts—but rarely even in the waiting room of consciousness.

All these thoughts and memories were news to me, unknown until I saw this transcript and others like it. I spent far more time with my mother's parents, and I was deeply influenced by the two of them in a million ways. Yet although all four grandparents appear in my dreams and sleep-onset thoughts often, my father's parents seem—to my great surprise—to appear more frequently by far. During some periods of my recent life, they have appeared every day.

Figuring out these sequences is a game of "know thyself." Element 4 includes two things I wanted very much but couldn't have. My Grandpa Sam was the most indulgent of grandfathers. This memory appears here not only because it fits geographically, but because it slips into the framework of his generosity. If he had known about the dog or the *Playboy*s I wanted, he would have done his best to get me at least one of each.

And Thus We Reach Dreaming

In dreaming, memory puts forward a sequence of hallucinated thoughts that are turned into one continuous hallucination, a story of sorts, by the conscious mind. One narrative episode turns into the next for no clear reason, as if the author of a novel were constantly receiving strange communications from Mars and were forced to work each one into his plot. It would work only if his listeners were amazingly gullible. But of course the sleeping self *is* amazingly gullible. We have only one mind. When we dream, one part is developing a sequence of hallucinated scenes, and another part is making them fit together into a story. So it's no surprise that we lack sufficient mind power for our *third* concurrent mental activity to be any more than passive and unreflective.

The dream is hallucinated, but not everything in it is imaginary. If you hear a car alarm in your dream, it might be hallucinated—or a car alarm might actually be howling in the streets.

An experimental psychologist summarizes the evidence from sleep labs—from experimental subjects deliberately awakened during periods when electronic monitors showed the body in the rapid-eye-movement (REM) state, which generates our more striking and memorable dreams. (The awakened subjects report what they have been dreaming and go back to sleep. At such times they probably envy their lab-rat colleagues; no one wakes *them* up.)

Compared to waking thought, dreams are "more visual, more perceptual, more affective, less thoughtlike."[6] More *visual and perceptual*: they express themselves in images and other sense impressions. Obvious but nonetheless crucial to bear in mind is that language is the abstract, up-spectrum way to express ourselves. Dreams are "less thoughtlike": unlike thoughts, they are not a response to, explanation of, or commentary on reality. They *are* reality, the world as our minds have made it.

They are, furthermore, "less concerned with contemporary life and more concerned with past life; more bizarre, implausible or novel . . . under less volitional control; accompanied by less awareness of the environment."[7] Obviously; but the "past life" is central.

A physiologist expresses it somewhat differently: dreams are "more emotional" than waking thought (treating "waking thought" as an undifferentiated blob). Compared to waking thought, dreams show "diminished self-awareness, diminished reality testing, poor memory, defective logic, and, most strikingly, the inability to maintain directed thought." Dreams are "perceptually intense . . . instinctive and emotional . . . hyper-associative."[8] (They show, that is, a large tendency to change the subject on the basis of free association from the last subject.) Notice the unreflective, collapsed self: "diminished self-awareness, diminished reality testing."

Losing touch with reality—withdrawing from external into mind-made reality—is a necessary, inevitable part of gliding down-spectrum: out of the world, into your mind. "Dissolving," writes Rilke, describes your sense of reality as you approach sleep. But from the opposite standpoint, of the remembered-yet-hidden past, "materializing" is the word. The immaterial, usually inaccessible sense of *your own past* starts to take shape just as your reality sense dissolves. Your *whole-you* sense materializes out of the mist; your sense of your life as a *real object in time* slowly materializes as you approach sleep through free flow, sleep-onset thought, and the outskirts of dreaming.

Moving down-spectrum, you gradually lose control of your thoughts, and your thought stream starts tampering with reality.

The dark side of consciousness, the low end of the spectrum, is poorly known by science—although it is a perennial obsession of mankind, one of the grand obsessions of human history.

Rilke, in the third "Duino Elegy," writes:

> *How he surrendered—. Loved.*
> *Loved his interior, his inner wilderness,*
> *this primal forest within him.*

It is the poet's privilege to speak of many things at the same time, but here a central topic is dreaming: "He was entangled in the farther-spreading tendrils" of his dream, of some "inner event," of his "inner wilderness," of "this primal forest within him"—*diesen Urwald.*

Falling asleep is a physical process associated with changes to your brain state. Ordinarily, "Is he asleep?" is a yes-or-no question. But one of the most important facts about thinking is the smoothness and steadiness of falling asleep. It is like walking down an inclined beach toward the water. At some point you are in the water; you are asleep. But you continue to walk slowly, gradually downhill. Your physical state has changed—you were dry, now you're wet—but the gradual downhill descent continues.

Proust begins the famous "overture" of his magnum opus with a beautiful description of continuity between awake and asleep: Sometimes,

> my eyes closed so quickly that I didn't have time to say to myself, "I am sleeping." And, half an hour later, the thought

that it was time to go to sleep woke me up: I wanted to put aside
the volume I believed I was still holding and blow out my light;
I had not ceased, in my sleep, to reflect on what I had just been
reading, but these reflections had taken a slightly peculiar turn.
(*À la recherche du temps perdu* [*In Search of Lost Time*])

We must recognize the continuity between sleepy and sleep*ing*
thought, as well as the physical transition that marks the start of
sleep. (Dreams soon after one has fallen asleep seem to be somewhat different from the "classical dreams" of REM sleep.)

Here's an example in which the dreamer is me:

[1] The dreamer is in London, deliberately formulating the
phrase "I will not go another day without stopping at the British
Library." He notices that the London streets are the same as
the Yale campus; the British Library is the Yale Law School. [2]
Somewhere on campus (or in London), the dreamer is watching
a fencing class for girls that involves weightlifting too; the
dreamer is on the floor of the gym, lifting high with his left hand
a 25-pound plastic-covered barbell weight. He hands it to the
(female) teacher, or class leader. [3] The dreamer is shagging
tennis balls, thrown or batted; the other players in this game
are staffers working for President John Kennedy. The dreamer
is catching the tennis balls left-handed and throwing them back
left-handed. He understands that he now has "excellent access"
to President Kennedy. [4] JFK appears, friendly and jovial,
wearing a formal morning suit or cutaway. [5] Waking suddenly,
the dreamer feels very happy; in his mind are the words, "Now I
have good access to JFK, and proper press credentials."

This is a continuous narrative instead of separate, disjoint hallucinations. But the real difference is small. Everyone knows that

the "plot" of a dream can take odd twists at any time. There is no more actual, logical continuity in a dream than in a series of sleep-onset thoughts.

This dream is a good example of a theme circle around a blocked emotion. The theme is a traumatic memory that is ordinarily banned from the thinker's consciousness. Twenty years before, he had suffered an injury that made it impossible for him to use his right hand in sports, and made sports difficult in general. He had been an enthusiastic (rather than talented) tennis player and occasional weight lifter in earlier life. He has come to admire President Kennedy for coping with constant pain and medication, but he dislikes (consciously) acknowledging his admiration, because he dislikes hero worship generally, because it is a cliché to admire JFK in particular, and because he is lukewarm on JFK's record as president. He never dwells consciously on JFK. Ordinarily, he also repels the traumatic injury from his conscious thinking—one result being that he dreams about it constantly.

Movies have a strong influence on the dream: one in which a woman learns to fence; another, an old movie we had just seen, in which men in formal cutaways man the front desk at a fancy hotel. (JFK wore formal morning clothes at his inauguration.)

Despite the somber theme, this was a (rare) happy dream—the happiness coming from the widely observed sense of gratitude and contentment when a person who is sick or hurt meets another with the same sort of problem—a person who will understand the accompanying inner reality.[9] The strange uses of journalist jargon—"good access," "proper press credentials"—are versions of a classical Freudian wish fulfilled: to talk to JFK. But there is more to it.

The dream shows Freudian wish fulfillment *and* an underlying need to tell ourselves the truth—*and* our quest to revisit the past.

When we are down-spectrum, we think in pictures. If we are

dreaming and hungry, we picture food. (We don't need to be asleep; it happens during normal, down-spectrum waking thought.) If we're down-spectrum and sexually hungry, we envision the right sort of object. If we're down-spectrum and want to be athletes again, or left-handed athletes, we envision an appropriate situation. In sleep-onset or dream thought, our envisioning takes the form of hallucinating

"Girls fencing" was floating around in the dreamer's mind because of a movie; combining that thought with his wish to be an athlete again, he shows off his re-created prowess to (naturally) a group of girls. Switching to tennis balls—well-loved objects long ago—he shows off, next, by joining an elite group of sporting young men. A White House staffer is elite by definition, none more so than a staffer in JFK's White House. This is all classic wish fulfillment, but it grows from the deeper and more basic force that lets banned memories slip into consciousness when you are down-spectrum and your guard is low.

"Access to JFK" means what it says, but it also means access to the whole glamorous, fabulous world of Washington and New York and other trend-making places—a world of people the dreamer often likes and admires. He had cut off his own access to that world, though—partly because of the injury, partly because of the gathering weight of obsessions, things he must do. And partly out of cranky reclusiveness and his failure to fit any standard, pre-stocked shape or size for academic intellectuals. Thus his wish to have "*proper* credentials"—in his case, impossible. His interests cut straight across half a dozen separate fields, like a biker roaming cross-country on his ridiculous Harley. Wherever he'd gone in life, he had been accused of (in effect) lacking *proper credentials*. Again, this dream shows classical wish fulfillment.

It also shows the usual reversion to the past—to the sporting past; to the days when the author was a sort of, tentative part of the "young elite" himself. Memories of JFK have many sources. Among them are early childhood memories of the man on TV, in picture magazines, and prominent in the conversation of adults and children alike.

Wish fulfillment dreams resemble sleep-onset thought, but dream thoughts show more distortion. Why? In part because dreams allow more time than isolated sleep-onset thoughts for a plot to build up and develop. (A big wave requires space and time to grow.) And a dream has narrative continuity. Narrative continuity is an expression of the same "continuity principle" that is expressed in our insistence on reading inner consciousness as reality when no outer reality is available. Likewise, believing that our experience of the world is continuous and *not* a series of separate bursts, we take the separate recollections of sleep-onset thought and throw a blanket of continuity, gently, over the whole.

When We're Dreaming, We Know It

According to Freud, the best dream observer we are ever likely to get, "I am driven to conclude that *throughout our whole sleeping state we know just as certainly that we are dreaming as we know that we are sleeping.*"[10] But we must add an important qualification: *we know this, but we usually ignore it.* When we watch a movie, we often experience exactly the emotions the film maker intended: anxiety, fear, excitement. Of course we know we are only watching a movie. But we set that fact aside. Dreams are not merely projected on a screen; they engulf us. If we set aside "this is only a movie," it's no surprise that our tendency to set aside "this is only a dream" is stronger; dreams seem more real than movies ever can.

In Sum

In the lower spectrum we revisit lost time and tell ourselves the truth. We know all sorts of truths we're not willing to tell anyone, including our conscious selves. But we need to tell someone, *somehow*, in a whisper, or we cannot rest. So we tell ourselves—and then forget. It's a compromise. We yearn to revisit the past, perhaps to be young again—at any rate, to go home. If this yearning were permanently unsatisfied, and there were no hope of anything different, our lives might always have that bitter, cynical edge they take on temporarily when some hope has collapsed or some project gone wrong. So we *do* revisit the past on the sly, in secret—a secret we keep from ourselves. We need the lower spectrum to do this for us.

Eight

Where It All Leads

Here I will discuss two questions that focus on applying the spectrum theory to the world at large. First: Does experience build confidence in the theory? Second: Does the theory help us understand psychological problems beyond the original one we attacked, the daily dynamics of the mind?

I will look at one fundamental problem and another that is peripheral but interesting. First, the mental development of children between infancy and the start of adolescence, at which point they are close to having adult minds (although they don't yet know what to do with them). Second, the meaning of visionaries in the modern world, and of the often-mentioned "spiritual state of mind" in which all nature seems connected. (By "visionary," I mean literally the seer who sees visions, not merely a big-picture thinker.)

The mental process of growing up resembles (as I have remarked) steady motion *up* the consciousness spectrum. It's as if infants were restricted to the very bottom of the spectrum, and in growing up, their mental romping grounds expand steadily at the high end. The mind is a sandbox with one end fixed in the low-focus world of sleep and dreams, the other moving slowly upward from the concrete into the abstract and logical.

Our mental romping room, this mental sandbox, expands steadily.

Eventually, the far end reaches the point where it will stay throughout adulthood, and the spectrum-based aspects of mental maturing are finished. The child's spectrum grows steadily longer until maturity, the way a twig grows. But of course the spectrum is not a mere straight path of a given length.

As I discussed in considering mental personalities (a Napoléon versus a von Neumann), one person's spectrum might be much roomier in some areas than someone else's. The lowest segment might take each one of us from sleep to the edge of daydreaming; but your spectrum might be a straight walk, whereas mine involves huge flights of steep stairs straight into the depths—and then a long climb back out again. By (metaphorically) scooping out the terrain in any part of the spectrum, we gain more depth and space in which to move. The spectrum is simple, but we mustn't make it too simple to account for the vast variation in mental personality.

Visionaries and spiritual states of mind give us a chance to consider atypical relations between a mind and the spectrum. I'll conclude that the spectrum is indeed a fundamental part of human psychology that helps us unravel some significant problems.

Children Developing

In standard terminology (where there is some disagreement, but not much), "infants" range up to eighteen months, "small children" to seven years. "Toddlers" are low-end small children. "Medium children" range from seven to twelve. The mental life of an infant is not easy to study, but I'll conclude this section with infants, who raise fascinating questions and possibilities. My main topic is small children. When we ask what's distinctive about them as compared to adults, we see a set of attributes associated with the lower spectrum. The process of maturing mentally, especially between eighteen months and seven years, suggests a slow march up-spectrum.

Children journey up and down the spectrum just as adults do; they are capable (just like adults) of different mental feats when they are fresh and wide-awake versus sleepy. But the *highest* focus a child can reach evidently *creeps upward* as he matures: the mental feats on which children improve as they mature are just the ones we associate with higher-focus thinking. Logic and analysis, abstract and systematic thought, sustained concentration, self-control—these tricks gradually enter a child's repertoire as he moves from infant to toddler to small child and onward. This is important, because it allows us to surmise new points about the spectrum and the value that I have called "focus level."

As we look back from the vantage point of adulthood, our memories begin somewhere between ages four and two. (Shakespeare took three years old as the starting point.) Cutting the road at that "somewhere" is a drawn curtain we can never see past. Freud called it "infantile amnesia." We recollect nothing earlier than that cutoff.

Growing up represents a *taming* (or breaking) of gigantic childhood powers of consuming the world, grabbing hold of it with both hands and chewing it over as you grab more. Children's curiosity has a purely physical side: childhood world-lust is a state in which looking over a fence, or down a pit, or over the heads of adults to a distant stage, or around a beckoning corner yields all but orgasmic satisfaction.

A small child's powers include aggressive, pushy consciousness expanding into every nook of outer and inner reality. A willingness to use *any* tool—to make up words or whole grammatical constructions, draw pictures, act, mime, whine to get points across; and the art of watching passively but ravenously. These small-child powers must all be cut down and sanded smooth to fit the needs of school and society. It all gets sorted out (we hope) just in time for the emotional chaos of puberty.

> ***Spectrum Development Principle:*** **The small child of about eighteen months starts with many low-focus mental attributes that he gradually sheds as he matures. In the maturing of small children, we see the development of logic and ideas of object constancy, of objective as well as subjective reality. We see aptness for concrete and visual thinking gradually fade, short attention spans and a focus on local instead of global knowledge gradually disappear, the gradual growth of mental discipline and self-control, the emergence of language, the abandonment of a passive "it must all be understood" view of information in favor of rejecting bad information and actively seeking out what is good. All these trends are consistent with motion away from lowest focus levels and toward higher spectrum points.**

Suppose It Were True

Suppose children's development did follow an up-spectrum route. What would the spectrum alone lead us to expect? If we project the adult's daily trip down-spectrum onto the development of small children, in reverse—the adult moves down-spectrum, the child up—we see striking similarities. We see spectrum changes echoed in slow motion in a child's maturing. In one case, centering on infancy, the spectrum suggests a strange and radical guess that, nonetheless, is supported by interesting data—and by William Wordsworth.

The timescale of the child's maturing is very different from that of the adult's day. Nor does a single rate of change govern the whole process of maturing. The child's intellectual achievements move gradually up-spectrum, but other lower-spectrum mental powers and habits might be filled in afterward. We expect the greatest changes between ages two and seven. But these ideas are deliberately vague. Much information is missing.

If small children do linger in consciousness zones corresponding to the lower reaches of the adult spectrum, short attention spans should be prominent in small children. After all, sustained attention to established tasks and problems disappears in the spectrum's lower half. (I'll use boldface to highlight what we would expect if the spectrum *did* predict the general shape of childhood development.)

Infants and small children do, indeed, have short attention spans. Studies (as well as most parents and teachers) confirm it.

Strongly related to short attention spans is the small child's tendency to concentrate on local neighborhoods, not on global or overall consistency. Children do not do well at coordinating and integrating many separate information sources either. It's only logical that they would build understanding of the world bottom-up, starting with details, not the big picture.

"Endless contradictions did not offend them," writes a kindergarten teacher; "the children did not demand consistency"[1]—meaning "global," or big-picture, consistency. Thus one of Piaget's most famous results: If you pour a fixed quantity of water from a wide, short glass into a tall, narrow one, have you increased the quantity of water? Small children, looking only at the height of the water level above the table, say yes. Older children take the big picture into account and say no.

For small children, in short, life is a series of chunks or incidents or events, each with its own coherence.

Now let's imagine a child who has finally made it to the rational, commonsense region of the spectrum, about halfway up. Most minds can achieve this level around age four at the earliest. The child awakens into a mental world where he can make himself pay attention, at least in short bursts. He draws on memory as an aid to thought. He uses informal, commonsense logic. His thinking is still concrete; in many cases it is visual. He does not use abstract words or ideas. His language is simple. But he is a rational thinker who can follow what's going on around him, listen to what he's told, pay attention. What do we expect from this child's mental life, if the spectrum idea is right?

We expect a steady rise in the highest attainable spectrum point until the age of sixteen or so. At sixteen, mental capacity is almost complete, and the rate of advance slows. *Does* the spectrum describe the facts of childhood development?

What would we mean if it did? A small child spends far more time than an adult down-spectrum simply because he *has* no up-spectrum. He's got nowhere to go, nowhere but down-spectrum to pass the time. (His colonization of the upper spectrum is steady but will take years.) So we expect low-spectrum phenomena to be *prominent*, emphatically present, in small children's lives.

Children also sleep more than adults, so they have fewer waking hours to dispose of. It could be that they sleep more, in part, because their spectrum is drastically compressed and may require less time to cover. Since the climbing rope is shorter, they hit sleep and the bottom sooner. Perhaps, then, the spectrum plays a part in a child's needing more sleep than an adult. We don't know.

In short, we should find lower-spectrum states strongly evident in small children, and upper-spectrum states *not* evident—but emerging gradually as months and years pass. Stronger command of language. Growing competence with abstractions—with numbers and measurement and arithmetic and time. Growing capacity to follow and construct logical arguments, and to solve real-world problems by reasoning.

Those traits do emerge as we expect them to. Language and number competence grows; at around seven, average children no longer need to count on their fingers. Abstractions become real to the mind. Between eleven and sixteen, children learn to think logically and to invent hypotheses.

What about low-spectrum states? Are they "emphatically present" in young children?

The small child who starts his day in the commonsense zone, and then moves down-spectrum as he loses focus and energy, *should* move downward through a mind-wandering phase, then a daydreaming or fantasy zone and a free-association zone, leading straight into sleep-onset thought and sleep itself. Does it happen?

Yes. Mind wandering, or not paying attention, is a house specialty with small children. Their conversation is noted for illogic and a tendency to veer around. Daydreaming is characteristic of small children. Sometimes they daydream like adults. But they also get caught up in "external" daydreams, where topics are fed to the passive mind not by memory but by the *outer* field of consciousness. Small children used to get lost in picture books. In modern times, many seem capable of watching TV or following a computer screen in a semi-trance. The result is an "out-of-body" daydream. Its contents come from the outer and not the inner world, but they are handled by a mind that is just as entranced, just as numb to everything else, as a normally daydreaming mind.

Thus, daydreaming *is* strongly evident. Free association and the sideways slide of thought are also childhood specialties, like mind wandering and short attention spans. "A common characteristic of young preschool children's conversation is *chaining*—that is, free-associating."[2] Children are good at thinking metaphorically, at understanding and inventing metaphors.[3]

Broadly speaking, things work out as expected.

Self-control dissolves dramatically as we move down-spectrum.

From the point where the adult mind starts to wander, we gradually lose conscious control over our own thinking. Daydreams and fantasy are sometimes chosen but sometimes impose themselves. Inspiration and creative insights hit unexpectedly; if we try to *make* them happen, we fail. Sleep-lab subjects "lost their control over the course of mentation" as they drifted off;[4] we have little control over dream thought.

Running things backward, we would expect small children to gain control steadily over their minds and behavior as they grow up.

They do. They gain the beginnings of self-control during early childhood.

Self-control requires alertness and energy. It also requires mental organization and discipline. It is acquired gradually in early childhood and is molded by many factors. One factor is, evidently, children's growing ability to internalize or say *to themselves* what adults tell them: stop talking, pay attention, stop fidgeting, stop tormenting your little brother. Also essential is the growing development of the idea of *self* as distinct from the rest of the world, which begins toward the end of infancy. (Small children are capable of such explanations as "the marble goes down the shoot because it wants to." The answer means, presumably, that the marble is more or less like me.)

We would expect the first self that the child develops to be unreflective and egoistic, like the dream self. Infants and very small children are wholly *un*-self-aware and unreflective. "The cognition of the infant is an entirely unreflective, practical, perceiving-and-doing sort of intelligence . . . not conceptual, self-aware, symbol-using."[5]

In the lower spectrum we would expect a strong bias toward inner versus outer field consciousness, as in down-spectrum daydreaming, fantasy, sleep-onset thought, and dreams. Needless to say, the child in the womb must be mainly inner-focused; there is not much going on outside. A newborn seems to engage the external world gradually: he sleeps (for one thing) between sixteen and twenty hours a day.

We would expect a strong bias toward concrete thinking, as in dreams; a bias toward visual thinking, as in dreams; the absence of language in the earliest period and its gradual introduction after that. Small children are no good at abstraction; they deal with the world *concretely*. The conversation of toddlers has a plastic quality, veering illogically from one topic to another. But each topic or subtopic is treated with reasonable narrative consistency.

Closely related is the observation that, of the many mental changes that accompany the infant's growth to small and medium child, the most important might be the transformation of a world of *seeming* to one of *being*, or (same thing) the crucial discovery that *seems* is not *is*. "In a kindergarten classroom," writes a teacher, "the appearance is as good as the deed."[6]

This is one way to interpret Piaget's celebrated idea of *object constancy*, or object permanence—an understanding that develops between roughly six and eighteen months, continuing to mature throughout early childhood. A physical object remains itself, although its surroundings and appearance might change—even if it temporarily vanishes. The child gradually reproduces mankind's momentous discovery of the *objective*. The discovery requires noticing the mind's power to model and keep track of the local environment—to build and update a mental microcosm of reality. It also requires the child's gradually turning his attention *outward*, from the engrossing inner field of consciousness to the outer field. Both developments imply that the child is learning to distinguish himself from the rest of the cosmos.

We would expect a bias toward visual thinking. We don't encourage children to develop their talent for visual thinking. Still, small children are rarely at a loss to express themselves with a crayon in hand. Even a child who draws badly (most children do, after all) will make an attempt to put down on paper the pictures he has in his mind. "Children under the age of seven are more responsive and attuned to nonverbal language—gestures and actions—then they are to words. They interpret words by relying on their understanding of the concrete circumstances around them."[7]

"Dante's attempt is to *make us see what he saw*" (T. S. Eliot, *Dante*; italics mine). The first duty of language after emergencies and propositions—the first duty of *descriptive* language—is describing images. Children show us mankind's original down-spectrum bias; Dante was sufficiently down-spectrum to see visions and think in images, but not far enough down to draw, instead of describe, the images he conceived.

William Blake, poet and painter, is the perfect uninhibited low-spectrum genius. His paintings are consistently awkward (and he *was* taught draftsmanship), but their originality, rhetorical force, and sheer energy make them compelling. Young children are uninhibited when they draw; no one has yet explained to them that they are no good at it. Mankind's neglect of drawing and painting as means of expression is one of the untold stories of culture making. It's true that a lack of suitable material for drawing is part of the story (although reeds and ink or something like them, and broken ceramic pots, have rarely been in short supply). But a deeper inhibition holds.

It's striking how much great twentieth-century art centers on intentionally childlike drawing: Matisse (in his early work and then his late cutouts), Miro, Vlaminck, Derain, and other fauves; Chagall, some Modigliani, some Franz Marc, some Bonnard; some Picasso (think of the 1950s); Dubuffet; many others. Great twentieth-century art *pushes deliberately down-spectrum*.

Storytelling is the down-spectrum counterpart of logic. Logic

has another kind of down-spectrum counterpart in "magical thinking," the idea that one can manipulate the physical world by thought alone. In magical stories, objects are created and destroyed, and their shapes and identities can change, when the right sort of thinking takes place. We find all these magical touches in dreams. Do we find them in early childhood waking thought?

Children love stories—both hearing and telling them. "She wants stories. I could imagine a woman who would make the child happy, filling her with tales from a past that really happened" (J. M. Coetzee, *In The Heart of the Country*). The two themes of stories and magic intertwine naturally: "Just thinking about magic was satisfying to the [kindergarten] children. If, in addition, they could talk about it, act it out, and put it into stories, they had strong feelings of contentment, of being in their own milieu."[8]

"Magical thinking," one species of illogic, is a prominent part of young childhood and continues strong into the middle period. In certain ways it is the opposite, or negation, of Piaget's object concept. But a child's tolerance for logical contradiction allows both to thrive. Anthropologists introduced the term "magical thinking" in trying to understand ancient and primitive peoples. Investigating a child's magical tendencies and where they come from is crucial to understanding childhood and infancy.

In magical thinking, things you only *imagine* become real. Objects, people, and places change identities. To small children, it is all perfectly natural. "Magic weaves in and out of everything [small] children say and do."[9]

A small child at a zoo saw a penguin and said "bird." As the penguin jumped into a pool, she followed it, examined it through the glass sides of the tank and said "fish."[10] Belief in a single object's unchanging identity increases dramatically between ages three and six.[11] In one study, three-year-olds petted a cat.[12] Then, while they watched the cat's tail end, its head was temporarily hidden and researchers installed a realistic-looking dog mask. The whole cleverly disguised cat was then revealed, and the children were asked whether the cat had changed into a dog. Many three-year-olds said yes. (See for yourself!) Six-year-olds were more likely to say that a cat *can't* change into a dog, period. An application of principle—of fixed identities in an objective world.

Unrealistic, illogical, or magical thought is a child's first guess at how things are. "As a child one can do without explanations. One does not demand that everything make sense" (J. M. Coetzee, *Summertime*). Children are famous for curiosity and asking questions. But they can make do without answers. They show more emotion than adults; but when they grow upset, it is not for lack of explanations. They can live without answers because, for one, they can always invent their own.

As children turn outward and make more contact with the outer world, they gradually tune the A-string of consciousness to the universal, uniform, world-standard pitch of logic. But over our lifetimes, the contrast between the "how does it work?" up-spectrum personality and the "look how beautiful!" down-spectrum variety remains marked. Some adults are restless if they don't understand the mechanism of whatever tool they use. Others retain the child's knack for awe, and thirst for sights and sounds for their own sakes.

Goethe's epic *Faust* is partly the story of a man's following the spectrum downward toward childhood, on a pilgrimage to the roots of consciousness, *away* from doing, analyzing, and understanding and *toward* pure being. If Faust is brought to say *Verweile doch, du bist so schön* ("Stay—pause—hold everything; you are so beautiful"), then the devil (Mephistopheles) has made him happy. He thereupon wins Faust's soul. Faust does indeed tell Gretchen, the perfect girlfriend, whom Mephistopheles has supplied for him, *Ein Blick von Dir, ein Wort mehr unterhält, als aller Weisheit dieser Welt*—"One glance from you, one word is worth more than all the wisdom of this world." One of the loveliest couplets ever composed. So Faust the "master doer" *does* come to rest for a moment on the soft shore of pure being, with translucent infinity lapping at his feet. Mephistopheles and Gretchen reduce him to deep and simple happiness (although his soul is rescued in the end).

Magical Reality

Before we leave magic, this strange jewel has one important facet that is widely neglected. But it is central to the child's mind—and to the soft, golden glow that childhood casts, so often, over the whole rest of life.

The infant's world *is* magical. For the infant, magic exists; magic is real. The central event in this world might be hunger and its satisfaction. The infant reacts to hunger instinctively by crying. Soon after, milk and mother (or a decent facsimile) materialize. (Not the inevitable but the usual case.) The child sucks, hunger disappears, satisfaction takes its place. *Magic!* The child has no world model to tell him that his crying *causes* milk to appear. Crying *happens* to the child, just as hunger does. Milk is what is needed to make you feel better, and at the crucial point, milk happens too. The infant's world *is* magical.

It's just because the infant *doesn't* understand causation, or the value of crying, that his magical thinking is realistic. There is nothing wonderful about a mother responding to her baby's demand for food. What *is* wonderful, what *is* awe-inspiring from the child's (or *any*) viewpoint, is the mother's simply appearing at the right moment and making things better. This is just what seems to happen if no ideas about communication, or cause and effect, clutter your worldview. For the small infant there is only hunger and the breast and sucking and feeling better. When a problem is solved, but we cannot say *how*, that is magic. The infant's enforced passivity, his careful watching (with a degree of attention and astonishment we can barely imagine), and his constant attempts to fit the puzzle pieces together, dispose him to see magic as a basic world force.

The child will come, in time, to see the mother as an agent who makes choices, not an impersonal force. But what *kind* of agent? Plainly, one who dominates and enfolds the child, and has the power to transform bad feelings into good. Some kind of goddess. (Infants, we assume, are born with no such concept but, gradually, invent it.) When the child feels bad, the mother-goddess appears and often makes him feel good again.

In short, an aging toddler, getting on toward two, looks back on a past that was magical and a present that *continues* to be magical in many ways (again, if things work just right but we can't imagine *how*, then they are magical). Eventually, the child realizes that his mother is a creature something like himself. But when he does, will he go back and touch up his memories of the benevolent goddess? His episodic memory is just barely emerging. His ideas about the magical past *could* be built into his mind as "basic principle" or "goddess concept," just as there comes to be an "object concept."

Freud's Oedipus complex, the "family romance" (as he calls it, ironically), is a dark reality in many lives. The small boy falls in love with his mother and sees the father as a rival; the small girl falls hard for her father and sees the mother as a rival. (The "Electra complex" is the girl's version.) It is all real. Not many memoirists choose to talk about it; it takes the brutal, beautiful honesty of a J. M. Coetzee to do that. "He does not want to have a father, or at least does not want a father who stays in the same house." *He* refers to the child Coetzee. "He denies and detests his father. . . . Since the day his father came back from war service they have battled each other in a second war" (*Boyhood*).

But we need to keep another primal myth in mind too—a joyful one; we might call it the "Wordsworth complex." It doesn't hold for everyone; neither does the Oedipus complex. But they are both common. Once upon a time, an all-powerful being was assigned to your personal care. Your own private goddess. The Wordsworth complex helps make adults see infancy as a sacred time.

Wordsworth believed that infancy is a stunningly luminous time when we grasp things we will never grasp again—although we might *remember* having grasped them, and that memory itself is precious. Wordsworth's mother died when he was eight. He was never close to his father. Naturally, his earliest life seemed like a dream. Yet the shape of his life and his gifts merely intensified for Wordsworth an experience that plays a part in many lives.

And in mentioning the mother-goddess myth as one origin of Wordsworth's belief in the sanctity of early childhood, I don't mean to diminish the spiritual significance of his vision. I only want to emphasize its solid foundation. He was not *inventing* the sublime glow that peeks out from behind the long-ago figure of the man's infant forerunner. For Wordsworth, and many others too, that glow was and remains real.

People of all religions and antireligions, and no religion, pray. The *Prayer Book for the Human Race*, that perennial best seller, includes familiar brief phrases to be muttered in emergencies. The most frequently used prayers in the book are probably the ones that express gratitude. We have powerful gratitude instincts—most of us. Partly they are superstitious, and obviously they are more. To whom are we praying?—those of us who are atheists, agnostics, nothings? For whom are we searching with our heartfelt *thank God*s, with our flashlights in the dark? For the mother-goddess of infancy? Too easy an answer.

We cast off too many other childhood habits too gleefully. I think that, in truth, nearly all of us *do* believe in God, although we don't realize it ourselves. The idea of God shocks and horrifies us. The original, most basic repressed idea of the modern psyche is our belief in God. The fact that we *do* believe proves nothing, except how much mind fashions change, and how much they matter. It's just *interesting*. Atheists, switch off your flashlights! Go back to bed. You only *thought* you heard a noise in the night.

Down-spectrum, memory is used concretely and associatively (we move freely along a chain of loosely related memories), not in abstract database-lookup style. At some point in late infancy comes a momentous conceptual discovery: the common time thread starts ticking, and it continues all the rest of one's life.

Infants have some sense of time long before the common time thread emerges as a constant background presence. They must distinguish the uncertainty of *now* from the certainty of *then*. They come to know when to expect important daily events. The light outside the window keeps changing in that certain way, and a vague idea of time opens a bud somewhere in the dense wildflower field of infant mind. The small shoot grows, presumably, into the idea of the unknown future threading the needle's eye of the present to emerge as the past. Then the common time thread, the ticking we always sense (although we lose it now and then)—the tape measure marked in time units unrolling beside us always—is unveiled. It becomes the master guide to all our memories.

Personality always matters, and sense of time varies from person to person like sense of space—better known as sense of direction. Some are born gifted; some are hopeless in these areas. Most fall in the middle. Nearly everyone can improve what he was born with if he likes. Sense of direction is the automatic plotting of our movements on an inner map. Landmarks and scenery can help, but true sense of direction is independent of all external markers.[13]

A strong sense of time is simpler: with the same independence of external markers, one always knows what time it is. Simple. Even if

I have no watch and haven't seen a clock or the sun for hours, I can usually bring the correct time into focus.

Some children are born with a strong sense of time; others are not. If you have one, it varies—like all our senses—with circumstances, with our sense of danger versus security. "Bobby's sense of time became acute. Without looking at his watch he could measure off quarter-hours" (V. S. Naipaul, *In a Free State*).

We all look back at our pasts with some feel for what happened five minutes ago, or an hour, a day, a week. Accuracy, and the rate at which accuracy declines with growing distance from *now*, depends on your time sense. But we all have a built-in timekeeper of some kind. Our sense of direction gets us around the landscape even if there are no clear markers, and our sense of time is our guide to memory in the absence of memorable incidents. Generally, the outer and inner fields of consciousness play symmetric roles. We look into the distance outside, or (within our minds) into the past. When we look into the past, the tape measure of our time sense guides our view.

We expect a powerful, enveloping (although usually not hallucinatory) imagination, as in fantasy and daydreams. We expect emotions to be vivid, and central to mental life (as they are in dreams and other low-spectrum states). We expect poor-quality memories that are hard to locate or recall (as in dreams and low-spectrum thought). We expect a passive, "dilate," permissive approach to stimuli from the outside world. Keats's perfectly dilate, low-spectrum openness to outside stimuli (his role being not to judge, but to feel and understand) is a model for what we might find in children.

Freud believed that infants hallucinate wish fulfillments. In effect, they experience brief waking dreams. A hungry child hallucinates the breast. But we have no way of testing this guess. It is relevant and suggestive that adult memories of early childhood are sometimes hallucinatory (as in Salaman memories)—and sometimes merely intense. Small children do have vivid imaginations. We all know it, but consider, for example, Singer and Singer's work on imaginary playmates.[14]

Some authorities go further—for example, if they are followers of the great visionary poet and painter William Blake. Blake believed, according to the author of the first standard biography, that seeing visions was "not an uncommon gift";[15] Blake "said to me that all children saw 'Visions' and the substance of what he added is that all men might see them but for worldliness or unbelief, which blinds the spiritual eye."[16]

When Blake wrote in his poetry "We see no Visions in the darksom air,"[17] he was saying that one fails to see visions when the *atmosphere* is wrong. An atmosphere (dark or bright) can be cultural as well as physical—as Blake well knew. *We* know that small children spend much of their time in the lower spectrum, on the long tidal flats that lead ultimately to sleep and dreams—where imagination is strong and bright. So it's not too strange for Blake to say that most children, in the *right surroundings*, would see visions routinely.

Children live in a rich imaginative world and are confronted with new information constantly, in huge quantities. The torrents they face reflect, partly, the lower-spectrum atmosphere in which they live; they have less control over their minds and the influx of information than adults do. And partly, they can't tell good information from bad, important from trivial—so they must try to make sense of it all.

Children *must* suspend judgment when torrents of new data overwhelm them. They *take it all in*. In the lower spectrum, the conscious mind is increasingly powerless against memory; it tries to make sense of each new dream scene that is pushed into consciousness. Recall that loss of control, the appearance in consciousness of *a thought we did not put there*, is the indispensable first step in falling asleep.

Infants are often, we assume, unable to judge even where one thing (anything) stops and the next starts. But adults, including the experts who run experiments, sometimes cannot follow small-child thinking.

Thus, one researcher found that small children are willing to reply to such nonsense questions as, "Is milk bigger than water?" This passive *refusal to reject* might be due (researchers speculate) to a child's eagerness to please adults, or to a lack of "metacognitive knowledge" that makes it possible to tell sense from nonsense. Far likelier, refusal to reject reflects a child's need to make sense of masses of new information hitting him in the chest like a swollen river; he is wading through as fast as he can go, as far as he can get. In a small child's world, "Is milk bigger than water?" is just another question to be analyzed as best one can.

Passive readiness to accept any and all new information and *reluctance to reject* information, ideas, or possibilities are crucial to a small child. Together, these two are strangely like the state of mind in which psychoanalysts are supposed to listen to patients. Passive readiness and openness suggest, too, the *sleeping* conscious mind's approach to whatever memory puts forward. *The sleeping adult mind makes the best of what it gets* and tries to assemble it all into a coherent narrative.

Adults, and children too, have nightmares. We are unable to reject painful or frightening thoughts when we are asleep. Nightmares—the mind is open to anything, even what is painful—echo the small child's willingness to make sense of "milk is bigger than water." The child's mind is open to anything, even questions that don't *seem* to make sense—and anything else that arises.

Dreaming and Waking

Suppose we apply the analogy between an adult's quick trip down-spectrum and a child's slow trip up-spectrum in the simplest, most direct way: the end of the adult's journey equals the start of the child's. For an adult, that final phase is part of one day; for a child, the *starting* phase is part of growing up. Suppose, then, that the child's eighteen-odd months of infancy roughly correspond to an adult's eight-odd hours of sleep and dreaming. The adult spends his *end* phase cycling between *conscious dreaming* and *unconsciousness*. The child should spend his *start* phase, infancy that is, cycling likewise between *conscious dreaming* and *unconsciousness*. Yet we are speaking of months, not hours, of the child's life. For much of that time he is *awake*. So how could he possibly spend his time cycling between conscious dreaming and unconsciousness?

While infants sleep between sixteen and twenty hours per day, Foulkes argues compellingly that infants and very small children *do not dream*; when they sleep, they are simply unconscious.[18] So the infant's periods of *unconsciousness* would be merely his *periods of sleep*.

Foulkes puts dreaming on the list of skills that small children must learn. The dreams of small children, he believes, develop gradually, like other mental skills. The earliest reported dreams are static images. Then come moving images. Then, finally, a "self" character emerges who takes part in the action. This emergence of a normal type of dream self doesn't happen, ordinarily, until past age seven.

Dreaming, Foulkes points out, is visual thinking in narrative form. Dreaming is *picturing* a story. We must learn how to do that. Are infants born capable of telling stories or of thinking coherently in pictures? There's no basis for thinking so. Foulkes also points out that infants have experienced too little to have amassed a good stock of memories as a basis for dreaming, even if they did remember things as we do—which they don't.

Now, if infants (theoretically) spend part of their days unconscious and the other part dreaming—and their sleep is, in reality, unconscious —where does the proposed "*dreaming*" component fit in?

Infants must dream *awake*. Many hints suggest they do.

Newborn consciousness does the sheer minimum to start. In *adult* life, that sheer minimum equals the role that consciousness plays in dreams. In dreaming, the conscious mind is pressed to the wall, and memory (or the unconscious) calls the shots. Memory feeds sequences of scenes into sleeping consciousness—memories of events, memories of *ideas* turned to visual form, sometimes a confusion or overlap of separate images. The memories might be recent, or they might have been summoned by a theme, often a blocked emotion. The conscious mind accepts these images, makes what it can of them, and strings them into a loosely structured narrative. The conscious mind, when asleep, is wholly unreflective—a mere intensely absorbed onlooker.

The most striking characteristic of dreams is that they are hallucinations. Despite Freud and Blake, there is no real basis for believing that infants hallucinate. But there is every reason to believe that infants' conscious experience *has an intensity, unexpectedness, magic, and mystery* that recall, as Wordsworth says, "the glory and the freshness of a dream" ("Ode: Intimations of Immorality").

If an adult merely exposes his awake but zero-focus, fully dilate mind in an entirely passive, open, Keats-like way to the external world, he might find himself wondering, as Keats did: "Do I wake or sleep?" ("Ode to a Nightingale"). He has just barely skirted a waking dream. And infants have far more to make their waking experience dreamlike.

Adults can barely grasp the surprise to an infant that is created by the unfolding scene. An adult always knows what to expect. Always knows where one object ends and the next starts. Can almost always *explain* what he sees. (The mother hears the child crying, which tells her he is hungry, which means he must be fed, and she feeds him. Where is the magic?) An adult has seen it all.

> *Whither is fled the visionary gleam?*
> *Where is it now, the glory and the dream?*
> ("Ode: Intimations of Immortality")

Wordsworth wants to know—looking back from adulthood. The infant's is a unique state of consciousness we experience once and then lose forever. But there are enough hints to let us reconstruct it, by doing psychological archaeology. (Freud invented and loved this image.)

Do infants dream awake? Consider an infant's mental life:

1. It is wholly unreflective.
2. It is dominated by sensations and emotions.
3. It is nearly empty of self-generated language and logic.
4. It is *magical*.
5. It seems to create no memories—or none that we can locate.
6. Wordsworth attributes to infant experience a huge, transcendent potency—a breathtakingly inspiring, transformative, yet *peaceful* excitement. "Infant sensibility," he writes, is the "great birthright of our being" (*Prelude*, Book 2).

Points one through five are all true of dreams too—the unreflective self, the dominance of visual and sensory elements, the illogic, the absence of newly created language, the *magic*; and of course, we remember dreams poorly.

Infants are perfect candidates for overconsciousness—consciousness burn, in which we are overwhelmed by sensory or emotional data and have no attention to spare on the recollections that form automatically within memory. Accordingly, these new memories are never hardened, never consolidated—and most can't survive.

The infant, awake, is alert and interested. But the texture of his experience is so wildly vivid that it crowds out analytic or reflective thought. The child is a mere entranced, passive, wide-eyed *watcher*; reality is always new, surprising, unpredictable. All this makes his ordinary daytime experience more like dreaming than waking.

Most memories of these earliest months can't survive. But some probably do. What would they be memories *of*? Presumably they would reproduce the general feel of infant consciousness, which (again presumably) would be a fine-textured blur in which one object or event melts into the next. The infant has yet to grasp fully that there *is* an outside world apart from his own mind.

But memories of continuously changing textures, one region blending smoothly into the next—memories like Monet's huge water lily paintings from the end of his life, where constellations of fine dabs create cosmic color clouds slowly evolving over yards of canvas—are *impossible to recollect* because they are impossible to grab hold of. No obvious search keys exist. Of course, those memories also have emotional content, are *steeped* in emotion. If we managed to grab on to such a memory by its emotional content, the rest of the memory would be a blur; there would be nothing else to it.

Spectrum Projection Principle: Adults asleep, at the bottom of the spectrum, spend part of their time dreaming and part unconscious. One quick description of the adult spectrum is *reality, dreaming, unconsciousness.* "Reality" covers everything from the top of the spectrum to the brink of dreaming sleep. There are reasons to guess that an infant's spectrum is the same without the top segment: *dreaming, unconsciousness.* In the abbreviated adult spectrum, the top zone is awake and the rest is asleep. In the infant's, same thing: the top zone [*dreaming*] is awake and the rest [*unconsciousness*] is asleep. The infant's awake dreaming isn't

the same as adult dreaming. But it is closer to adult
dreaming than to adult waking.

What is it for an infant to dream awake? Perhaps it's like confronting great crashing breakers that dominate the mind and leave the observer transfixed. The subject matter of these "dreams" comes from outer instead of inner consciousness—from perception instead of memory; from *now* instead of then. But the dreamer's attention is glued to the scene before him, whatever it is; and his "self" barely exists.

Coleridge's "Frost at Midnight" includes a fascinating reflection on the closeness that children experience between daydreaming and real dreaming, and the intermingling of awake and asleep. As a homesick child at school in London, "With unclosed lids, already had I dreamt / Of my sweet birthplace," far away. The child Coleridge would continue daydreaming:

> *till the soothing things I dreamt,*
> *Lulled me to sleep, and sleep prolonged my dreams!*

The Spiritualist and the Visionary

We don't, ordinarily, value a talent for *being* as we value a gift for doing. People whose best gift is simply to be who they are, having *their* personalities and no others, instead of some art or craft or knack, are sometimes admired and respected and loved within their own worlds—and that, after all, is what counts. But their voices rarely carry very far.

Some people do have spiritual gifts, though: they can *be* a certain way, and feel part of something larger than themselves. This is not the same as religious yearning or feeling, although a spiritual gift can sometimes make a person more religious. In itself, spirituality is a talent like perfect pitch, or the ability to draw likenesses. They are all mere psycho-physiological quirks. Yet they grant entry (which you might use or ignore) to a deeper world—of music, art, metaphysics.

Perfect pitch is a retentive memory for pitch. You almost certainly remember middle C from the last time you heard it; you don't have to look it up. A good ear makes it easier to hear music with analytic understanding. As for the spiritual gift, it has many possible consequences but centers on one particular feeling. This one feeling can be grasped in terms of the lower spectrum, where heightened emotions make it easy to summon recollections using shared emotion as bait, and to surf the long, subtle seams of memory—the "isofeels," the many recollections that *feel* the same and must therefore be, somehow, related.

At the start of "Mont Blanc," Shelley writes:

The everlasting universe of things
Flows through the mind, and rolls its rapid waves,
Now dark, now glittering

Everything flows together in review, thing by thing, a single wave. In *The Brothers Karamazov* of Dostoyevsky, Alyosha "began quietly praying, but he soon felt that he was praying almost mechanically. Fragments of thought floated through his soul, flashed like stars and went out again at once; to be succeeded by others. But yet there was reigning in his soul a sense of the wholeness of things."

Wordsworth writes in "Tintern Abbey":

And I have felt

· · ·

A motion and a spirit, that impels

All thinking things, all objects of all thought,
And rolls through all things.

Now let's define. The spiritual gift allows a person to feel (not deduce or decide) a transcendent unity among far-flung objects and events. This experience of cosmic unity often (though not always) suggests one creator who stands outside his creation. A feeling of cosmic unity can make a person feel *outside of*—over and against—creation. The connection between spiritual feelings and creativity is obvious: if creativity centers on discovering new analogies—connecting a pair of different-seeming things and drawing conclusions—spirituality centers on making long chains of such connections. The creative person discovers something. The spiritually minded person *experiences* something: the unity of many people, objects, or events—or of everything in the cosmos.

This is no belief in underlying unity; it is the direct *experience* of underlying unity—a far more formidable thing. Cosmic unity becomes an emotion, "felt in the blood, and felt along the heart" (Wordsworth, "Tintern Abbey"). It is important that we understand the spiritual gift—first, because of its intrinsic interest (and association with interesting people); second, because it is a major force behind some religious experiences, especially in the mystical strains of Judaism and Christianity.

This feeling of pervasive unity has several possible consequences:

1. It shows some spiritualists that there is more to life than one can measure or objectively classify. After all, the many objects caught up in that unifying mental tide don't *seem* related at all.
2. It gives some spiritualists a sense of duty to all mankind, or to all living things. They might seem radically different from each other and ourselves, but one *feels* that they are parts of a great unity.

3. It makes some spiritualists feel the presence of God. This felt presence of God is (again) not at all a *belief.* In principle, one can be argued out of a belief, but never out of a feeling.

We know that two persons, scenes, or anythings that *seem* very different can evoke the same, or almost the same, emotion. And emotional response isn't random or arbitrary. We don't know how it works; we assume that it works differently for everyone. In any case, *something* accounts for two different things evoking the same emotional response—a particular attribute that the two things share, though it might lie deeper than can be detected in plain thought or described in words. You walk into a hotel room you've never seen before. Something reminds you of your kindergarten teacher and your first day of school. That something might be significant or trivial; anyway, it exists. Two things producing the same emotional response in the same person must share *something.*

When two different things are tagged with the same emotion, it becomes possible to *recall* the second when we encounter the first. (Assuming we can recall the emotion that the second thing produced. Sometimes we do.) If I encounter A and am reminded of B, my mind is proposing an analogy. Since A reminds me of B, it's natural to suggest an analogy between A and B.

Emotions aren't just remarkably powerful summaries or abstractions. They are *content-transcending* abstractions—*meaning-transcending.* The same emotion might be produced by two entirely different-seeming things. An emotional response might (often does) reflect a particular surface attribute, but it might also reflect a deep attribute that has nothing to do with the thing's meaning or significance in any obvious way. (A certain person's face with a certain smile, and a single close-fitting quartet of small Virginia creeper leaves, orange in the fall, picked out against a dark trunk by late-afternoon sun, might cause the same flavor of joy.) The fact that emotional response can transcend (or be independent of) content or meaning gives it the power to inspire surprising new analogies.

Because emotional responses transcend content or meaning, they make it possible to move "sideways" freely from thought to thought: thinking about trees in forests, I find myself thinking about high-rise buildings in cities. A small, subtle shift in emotion might lead me onward to flower beds in a Victorian-style garden, then to a movie about the painter Renoir, and a painting at a certain museum, and a girl I once knew, and a shop in New Haven, and the elderly owner behind the cash register, and a swimming pool in Westchester, and on and on. The spiritually minded person makes the subtle adjustments in emotion that allow him to ride this current indefinitely, like a bird gliding along a boundary between air masses, or a surfer making the delicate, constant adjustments to his position that let him ride a wave's inner curl forever—or so it seems.

The ability to sustain this mental swoop is the gift of spirituality. The meaning-transcending nature of the connections the thinker follows creates the impression that he is surfing *on* rather than *in* meaning. Important! He *feels* unity, insofar as he passes from one thing to the very different next and the next and the next while *feeling* just one emotion—or one slowly, subtly varying emotion—as he goes.

All those dramatically different persons, scenes, events, objects make him feel *almost* the same way; he glides from one end of the world to the other on *one* gently varying emotion—and there is the feeling of unity. That feeling transcends the meaning or nature of *all* these separate things, and so he *feels* himself transcending meaning. Once again, Shelley:

The everlasting universe of things
Flows through the mind, and rolls its rapid waves,
Now dark, now glittering

But a spiritual gift isn't merely the ability to sustain long associative chains. It is the turn of mind that makes one *aware* of these thoughts. It's normal to free-associate as we slide toward sleep. As we move down-spectrum, our thinking grows increasing more likely to slip sideways—from one topic to a completely different one, and another and another—rather than burrowing deeper along logical lines, sticking to *one* topic. But ordinarily, by the time this happens we are sufficiently far gone toward sleep, toward pure being, that we experience our mental states unreflectingly and recall little or nothing afterward.

By *reflecting* on the free-associative chain, the spiritually minded thinker makes himself aware of what is happening in his mind. This gives him a chance to *feel* the encompassing unity. One needn't recall each (or *any*) of a free-flowing sequence of thoughts. One need only recall the feeling of the experience as a whole—of gliding smoothly from one thought to another and another *above* the level of meaning, where the overarching unity of all persons, creatures, or things makes itself felt. *Felt.*

The Sanctity of Long Ago

For Wordsworth, the past is sacred as a fire is warm: the past *feels* sacred as you approach in your mind. He refers to his own past. But other poets, and thinkers of all sorts, have felt warm sanctity in ancient history. The past nearly always attracts us—if we are not time-blind innocents who cannot see ten minutes *into* the past—because we can take it in our palms and see it whole. We like, we *need*, to control our own worlds—although we rarely can. But the past seems controllable, therefore safe. It's an illusion, but harmless. We know it to be an illusion as we know a movie is a movie and a dream is a dream, but we set these facts aside. And it is easy to guess that we project the sanctity of our own early childhoods onto history at large.

But not so fast. The same stretch of spectrum that separates early childhood from adulthood separates ancient civilizations from our own.

The lower spectrum zones make spirituality possible by allowing certain people an experience that doesn't happen up-spectrum. They feel the unity of the cosmos. When poets tell us about the sanctity of ancient times, they are referring to eras when people lived lower on the spectrum than most moderns do, and when spiritually minded people were more common than they are nowadays.

Jews, Christians, and the faithful of many other religions believe in a spiritual golden age long ago. If we dismiss their beliefs, or put them down to a mistaken admiration (subtle as Stonehenge) for old things we don't understand, we are missing something important. Let's glance briefly at a couple of poets instead.

Blake's *Songs of Innocence and of Experience* deals with two worlds in tension, coexistent in history *and* within each soul—innocence and experience—but also an "innocent" *married* world versus a world of sexual longing and passion. ("Sooner murder an infant in its cradle than nurse unacted desires," Blake wrote, as I've mentioned, in the *Marriage of Heaven and Hell*. Yet he was himself a gentle man with a happy marriage.)

In the *Songs of Innocence*:

Piper pipe that song again—
So I piped, he wept to hear.

Such, such were the joys.
When we all girls & boys,
In our youth-time were seen,
On the Ecchoing Green.

But the *Songs of Experience* speaks of departed sanctity:

> *Hear the voice of the Bard!*
> *Who Present, Past, & Future, sees;*
> *Whose ears have heard,*
> *The Holy Word,*
> *That walk'd among the ancient trees.*

The Bard, insofar as he knows the past, has heard the holy word, which walked among the trees of antiquity. The ancient world seems thick with spirit. Feelings leapt from man to man, and one man's direct spiritual experience of cosmic unity was easy to communicate—though not (not mainly) in words.

In "Ode to Psyche," Keats calls to the pagan goddess across time:

> *O brightest! though too late for antique vows,*
> *Too, too late for the fond believing lyre,*
> *When holy were the haunted forest boughs,*
> *Holy the air, the water, and the fire.*

Sanctity long ago was in the very air, the water, and the fire.

The spiritually inspiring feeling of cosmic unity would never have been widespread, in antiquity or any other time. But surely it was less uncommon long ago than it is today. It might even have been fairly familiar at second hand.

Spectrum theory suggests that the average person in such societies was "more emotional" than we are. Such people were readier to entertain emotion when it appeared, in either a recollection or an immediate reaction (to a recollection or idea, or to the outer world). Such people would have made society a better conducting medium for emotion than we are, we cold fish. We are well insulated in our mental worlds of abstract thought and heavily suppressed, harshly sat-upon feelings. Emotions would have spread more readily in lower-focus societies, flashing across the gaps between person and person. *Spiritual* feeling might have spread far in a society that was obsessed with the sacred and the spiritual.

We know that the body speaks its feelings. People read each other's feelings with no words to help. Sometimes they feel each other's feelings. Members of ancient societies (not *any* old society, but certain ones) would have been more aware of each other than we are today. They would have been more "plugged into" each other, more apt to feel each other's feelings.

To learn how to communicate with their fellow human beings, young people must turn off Facebook, shut down their computers, and look people in the eye, listen to their voices, watch their gestures. They must look for subtleties and ponder their meanings. They must learn to read, not words (which are easy) but people—and *that* requires a whole childhood and adolescence to learn. Some people never manage it, although they try; this sort of reading, the important kind, requires intelligence and talent, not just a few years' dogged practice. By allowing children to play with computers when they should be dealing with each other face-to-face, we are damaging the most important learning process of their lives.

The Visionarics

Seeing visions is a different low-spectrum effect. This is the zone of mental halfwayness. We are caught between awake and asleep. This "borderground" as Edgar Allan Poe calls it (*Marginalia*) is one of the most fruitful in the spectrum for the intense imagining that creates poetry, vivid language, and striking mental experience. We can feel the closeness of sleep—of sleep-onset thought and dreams. And just as we can pull mental energy out of the up-spectrum zone when we need it, although we are *not* up-spectrum—as we might catch a drifting rowboat and haul it out of the water onto the beach—we can pull parts of dreamlike thought out of the sea onto the low-spectrum borderground.

Poe was a poet with a deeply musical ear. He describes the borderground as "where the confines of the waking world blend with those of the world of dreams." This borderground is (naturally) "upon the very brink of sleep." For Poe, it is a mental realm unlike any other; it's not surprising that he should emphasize its sensual richness—here we are deep into the world of being, of sensation and emotion versus rational thought.

"It is as if the five senses were supplanted by five myriad others alien to mortality." We are overwhelmed by sensation. As we withdraw into mind, into the inner field of consciousness, we are no longer directly connected (via the body) to the outside world. Now our thoughts look *through* memory (as if through stained glass) at the outside world, and their texture has changed. We have lost the sense of direct linkage to the world outside.

Coleridge was intrigued by the state of "half-awake & half-asleep"; he composed "Kubla Khan"—he called it "a psychological curiosity"—after he had slept and dreamt in an opium haze. He conceived the poem in his dream and set to writing it down as soon as he awakened. But someone interrupted him midway through, and he forgot the rest.

Büchner's slightly insane Lenz tells himself, "If I could only decide, whether I am dreaming or awake."

Shakespeare's Macbeth is a man who lives in the borderground. He experiences inner reality as vividly as outer reality, or more vividly—as if his two fields of consciousness had got confused. Macbeth "stands in doubt between the world of reality and the world of fancy," writes Hazlitt; he lives "in a waking dream" (*Characters of Shakespear's [sic] Plays*).

Kafka alone ventures all the way to the spectrum's bottom. That Kafka's greatest works make sense only as *transcribed dreams*—that they obey dream form strictly—is so obvious and important that many critics don't even bother to say it. (The best take it for granted or discuss it explicitly. Some seem unaware of it.) Kafka composed dreams and wrote them down. He used the dream form with the freedom any great artist brings to a new and original structure. But his starting point is clear.

Kafka himself spoke of "the dream-like life of my soul." His work is so true to dreaming that, beyond a certain point, we stop using dreams to understand Kafka and start using Kafka to understand dreams. Understanding Kafka's genre can't explain his complex works; it is merely the indispensable first step—like getting the clef and key signature right when you look at music. If you miss this point about Kafka, you miss everything.

In most of his stories, the mood is that vague but persistent unhappiness, or "dysphoria," that psychologists find so often in dreams. One scene transforms itself into another; characters enter, disappear, and are forgotten; the spotlight wanders as new facts replace old ones; scenes elaborate themselves *just insofar as the narrator turns his attention their way*—a central facet of his work, and of dreaming. The main character's goals and motives change (again, a central attribute of dreams), and the story slips and slides forward as if it were walking on sheet ice in a gale. Just as dreams do.

Often Kafka's narrator can't quite make out what is happening: "The verger started pointing in some vaguely indicated direction" (*The Trial*); "a noiseless, almost shadowy woman pushed forward a chair" (*The Castle*). Bizarre events call forth exactly the unreflective, incurious, matter-of-fact response that is the very essence of the unreflective dream self: One morning, Gregor Samsa "found himself transformed into a gigantic insect" (*The Metamorphosis*).

"Nobody knew where to bury him for a while, but in the end they buried him here"—under a table in a tea shop ("In the Penal Colony"). Any Kafka reader will recognize instantly that this catalog could go on for a hundred pages. *Dream is as plainly Kafka's chosen form as the five-act play is Shakespeare's and the three-volume novel Jane Austen's.*

In Sum

Often the first step is obvious: start the car; write one chapter; ask her out. Then the complications set in. A normal adult spectrum cannot match up better than *approximately* with a child's gradual development. But it's easy to picture the unfolding, blossoming mind of a child resembling (in reverse) the gradual withdrawal of the adult mind from outer reality. This is a matter of amassing and then artfully charting large numbers of data points.

Spectrum *hermeneutics* are different. The spectrum is ready right now to make practical contributions to the understanding of the Hebrew Bible, and other ancient literature, and literature and culture in general. Yes, reading is out of fashion nowadays. But the wheel will turn. Some scholars, including Professor Ruth Morse,[19] believe the turning has already begun.

Modern-day relations between science and religion are all wrong. "Science" has no more right to pontificate about religion than it does about field hockey or dog shows. Science does have an unmatched record of producing useful tools. It should produce intellectual tools for the use of religious thinkers as it does for so many other fields, deliver them, and keep its adolescent wisecracks to itself.

Nothing is sadder than an eminent thinker's making a fool of himself by explaining or denouncing things he doesn't understand. Of course, there is no reason a scientist shouldn't become an expert on religion or atheism and speak of those things with authority. The problems arise only when scientists believe themselves qualified, just by virtue of *being* scientists, to tell the rest of us how to think and what to believe.

Is the spectrum one of those tools that could help religious thinkers understand their topics better? I hope so, and in this book I have tried to show how that might work.

Nine

Conclusions

The Basic Points

I. THE MIND'S STRUCTURE

(a) The mind has two separate regions: conscious mind and memory. Memory is equivalently "unconscious mind."

(b) Conscious mind deals only with *now*; memory deals only with *then*. The only instant of time we can ever experience is *now*. When we reexperience the past in the hallucinations of sleep-onset thought and dreaming, the past comes to *now*, not vice versa.

(c) Conscious mind is a spectrum from pure *thinking about* to pure *feeling*. A feeling is a sensation, emotion, or mood.

(d) Conscious mind is a spectrum from pure *acting* to pure *being*—from pure mental action to pure states of being. "Being" means participating in, specifically *experiencing*, the state of an object—in particular a human being: body and brain. The mind *experiences* or *feels* the state of the object in which it participates. Happiness, sadness, moodiness, pain, sleepiness, joy, and so on are the results.

(e) Feelings are inflections of consciousness. Feelings cannot be unconscious.

(f) Being is not computable. That is, software cannot yield *being* as its result or output. Being is the sentient element of an *object in a particular state*. Being *presupposes* a physical object in some state. "French fried" is not computable. Why not? Because it is a particular state of a physical object. It describes a physical phenomenon—as do "rusty," "broken," "feathered," "transparent," "ferrous," and so on. "French fried" means that a *certain part of the universe* is in *some particular state* versus some other state. It reflects the physical nature of an object and its recent history. It is not a mapping from numbers to numbers (that is, it is not a mathematical function) and can't be created by any such mapping. Happiness is not computable. Feelings are not computable. Thinking about (as a process yielding a result or outcome, an *action*) is, in many cases, computable; states of being are intrinsically not computable. A mind entails thinking about *and* being. Computationalism, therefore, is wrong. The analogy that likens the mind to software and the brain to a digital computer is deeply misleading.

(g) Thoughts are independent of the body. Feelings are part of the body.

(h) Thoughts *must* be about something. Feelings *cannot* be about anything. All thoughts are intentional states, and all feelings (properly so called) are nonintentional states.

(i) A mind requires a body and a brain.

(j) In descending the spectrum, we pull out of the world and into the mind.

(k) Logic and storytelling are counterparts—different ways to account for reality. Logic assumes that all main actors are logical actors and offers logical explanations of their actions. Storytelling assumes that all main actors are human actors and offers psychological explanations of their actions.

(l) Hallucinations are the logical end point of the increasingly attention-monopolizing, external-reality-excluding series of mental states starting with mind wandering and daydreaming.

(m) Thoughts are expressed deliberately in language; feelings are expressed involuntarily (in the first instance) by the state of the body. "State of the body" means facial expressions, tone of voice, gestures, and so on. Feelings can be expressed in language, but language is secondary to body state in communicating feelings.

(n) We understand each other's thoughts on the basis of understanding language. We understand each other's feelings on the basis of implicit pointing out: one person points out a feeling, known to another person, that corresponds to the first person's feeling. The implicit pointing out is done by facial expressions, gestures, positions and attitudes of the body, and so on. We can know another person's thought even if it has never occurred to us. We can't know another person's feeling unless we have experienced it (or something much like it) ourselves. Thought can be communicated only in language. Feelings are ordinarily communicated without language.

2. MEMORY

(a) Memory's spectrum of functions ranges from supplying abstract information, through returning specific recollections, to (finally) supplying specific recollections or groups of recollections that are transformed into hallucinations.

(b) Memory *creates* and *retrieves recollections*, or information yielded by recollections, and allows recollections that are similar to settle (compress, or compact), thereby yielding abstractions (templates, or schemata) and tidying up memory by replacing many entries with one and forgetting unimportant details.

(c) Both the creation and the retrieval of recollections are ordinarily unconscious—but they respond to events in the conscious mind.

(d) Compaction (template formation) is ordinarily unconscious— and independent of events in the conscious mind.

(e) There is only one kind of template, rule, or archetype, whether it deals with an object in space or an event in time.

3. FEELING

(a) Emotion runs the mind. Up-spectrum, we choose the goal or topic of our thoughts on the basis of emotion. Down-spectrum, explicit emotion becomes increasingly prominent in determining state of mind, in recollection, and in associative chaining.

(b) Emotion is our most powerful *essence summarizer,* or *content-independent summarizer,* of a complex, multipart scene or recollection. It is abstract insofar as an emotional summary depends only *indirectly* on the contents of the summarized scene or recollection. One emotion can serve as the summary for seemingly very different scenes or recollections.

4. CONSCIOUSNESS

(a) Consciousness has two fields, outer and inner; we switch gradually from outer to inner as we move down-spectrum. Perceptions define the outer field; recollections and ideas, the inner field.

(b) Consciousness *always* has a "quality," *sometimes* a "target."

(c) Feeling is the *quality* of consciousness. Feeling is the current value of the quality of consciousness.

(d) Feelings are *inflections* of consciousness. Feelings define the quality of consciousness.

(e) When we are *just* happy, we are conscious of nothing. Yet we *are* conscious. When we are just happy, our quality of consciousness is "happy" and the target of consciousness is null. If we *think about* ourselves being happy, then the target of consciousness becomes "our own happiness."

(f) Consciousness is the sight and the seer simultaneously—the observed and the observer.

5. PARADOXICAL EXPERIENCE

(a) An experience is "paradoxical" if it happens to us but creates no memory, so it never *happened* to us. We experience it as an event in the *present* but never the past. Informally, paradoxical experience falls out of our lives. If we can't remember it, it was no part of our life experience.

(b) A paradoxical experience is not part of our lives.

(c) A paradoxical experience of X is paradoxical only because of the *nature* of X (only because we were in the right spectrum region for X—always low-spectrum).

6. DREAMS

(a) Dreaming is remembering. A dream is the conscious mind's attempt to understand the unsolicited output of memory. Blocked or frustrated emotions are especially important in determining memory output at bottom focus.

(b) In the right circumstances, dreams are good predictors of the future. They are good predictors when they expose attitudes or facts we know but are unwilling to allow into waking consciousness.

(c) Infants dream *awake*—but their dreams are fed by outer rather than inner consciousness, by *now* instead of then.

(d) A dream is a theme circle, a collection of scenes or incidents all attracted by the same memory cue.

7. CASUAL OBSERVATIONS

(a) The most widely repressed public emotion in modern Western life is belief in God. God is not "dead." In Western society, God is a widespread, powerfully repressed belief. This observation has nothing to do with the existence or nonexistence of God.

(b) The most widely repressed private emotion in modern life is homesickness for homes that no longer exist—are lost in the past, or in some other way: lost-home-sickness.

(c) If we confront our dreams and sleep-onset thoughts, we will learn what's on our minds.

(d) Every night, our minds tell us who we are. Every morning we forget.

Where Do We Go Now?

Forgetting protects us from the painful bouts of remembering, self-accusation, and plain honesty that flourish in the spectrum's lower reaches. But we might accept the danger if we knew how to expand our lives into these lush, vivid lower regions where the past is hiding. We gain greatly when we learn to see things we never used to notice before. That's what happens when we learn to recognize trees or flowers or birds or butterflies, or buildings and architects, for that matter. Knowing each one's identity is never the real achievement. The achievement is *seeing* better—looking carefully at things we barely glanced at before.

All sorts of people have invested time in studying the worlds around them. Most of us invest almost none in knowing ourselves. This book could almost have been a *Field Guide to the Mind from Inside*, but it's far harder to make ourselves at home in the lower spectrum than in fields and forests. Nature means for us *not* to remember the lower spectrum.

Still, anyone who chooses can learn something about the depths of his own mind, visited and forgotten every day. One can learn to step back—just a bit, now and then—from the sleepy states in which wandering thoughts turn into a free-flow of recollections. Stepping back to reflect is a habit one can learn like any other. To step back from sleep-onset thoughts or dreaming is harder but not impossible. The way to start such a project, though, is probably to rouse yourself (you can't *wake* yourself because you're not sleeping) during sleep-onset thought and look around. You might be surprised. Of course, just figuring out when you are apt to be in this state is a project in knowing yourself.

Different views of mind have different consequences: therapeutic, scientific, philosophical. The spectrum view cries out for research in physiology and psychology. But its immediate uses are hermeneutic. How do we read literature, culture, and ourselves, our own lives? The spectrum broadens and sharpens our view of mind. It suggests many topics for study: memory as an information source versus a source of anecdotes, analogies, and alternate realities; theme-circling narrative as a system in its own right; the dream as a literary form; the tremendous encoding and summarizing power of emotion.

That emotion coding (the powerful information-processing capacity that is so different from digital computing) is inherent to the human mind is suggestive and important. People differ greatly in their talent for emotion coding—in the robustness, accuracy, and nuanced precision of their performance—and the differences must be studied and understood. (Accurate emotional response, anticipating such responses and allowing for them in one's own plans, are focuses of Peter Salovey's Center for Emotional Intelligence at Yale.)

Access to the relatively creative, bold, improvisational, unconstrained precincts of your personality depends on normal access to your lower spectrum, normal lowering of your mental focus. We all know people who don't seem to *have* that normal access— who seem stuck in the emotionally limited, abstraction-rich upper spectrum not because of boundless mental energy or joy in mathematics or abstract thought, but only because something keeps them up-spectrum when they want and need to move down. We know nothing about the meaning of such phenomena. Could focus-lowering therapies—talking cures, drugs—be useful? We are nowhere within shouting distance of an answer.

There are many other topics on this list of barely scratched surfaces. In some cases we lack even the proper words for asking our questions.

We must know how the body changes during down-trips versus up-trips along the spectrum. We must understand mental focus in physical terms. We require this knowledge to satisfy our curiosity, but also to help in treating spectrum-related illness, especially insomnia, which can be devastating and is not always taken seriously. We want psychological data from many detailed interviews to fill out the bare outlines of the spectrum presented here.

The spectrum underlines how much of our lives we miss, insofar as we can't remember them. Most of our dreams, of course, and nearly all the mental life that leads up to dreaming, and those strange dreamlike moments during the day—the common sort that Börne discusses and the (probably) rare Salaman type. We might choose *not* to miss them because these moments include passages of childhood and youth that we would like to reexperience—even though they might not be pleasant in themselves.

We could teach ourselves not to miss this experience if we chose to. We are capable of bending mind to will in all sorts of ways. We acquire and lose neurotic symptoms; master new languages; learn to hear, with understanding, a sonata-form movement or an engine or the crack of a baseball bat. There is reason to believe that an older society's tolerant interest in visions encouraged the emergence of visionaries.

———

There is a powerful tendency in philosophy and cognitive science to ignore sleep and dreams, or at least to toss them into a separate category—a bargain basement for intractable merchandise. By neatly shearing off the bottom of the spectrum, the shear wielders create the illusion that there *is* no spectrum, no change over the course of the day. Just *thought*, period; *mind*, period—two great gray static lumps. Granted, it is harder to understand a dynamic, time-varying system than a static one—*but that is the human mind!*

Mind researchers do the same thing when they ignore childhood thought. Shearing off childhood leaves, again, the illusion of one unchanging slab of thought: by age twelve, children perform much like adults on many scales. But again this result is false to nature. In nature, the mind is dynamic: it changes slowly over time as we mature, quickly as we move through the day. That is nature as we find it.

———

The arc of mental life between doing and being, thinking and feeling, is the place where we live. It has boundaries. You can see one end from the other. But the field in between is infinite because it contains a continuum of possibilities, like the space between 0 and 1.

It's easy to see that wide-awake energy yields a different mental life than tiredness does, or drowsiness or sleep. Easy to see, also, that the short ride from the first stop to the last is no simple decline. Metaphor and creativity, and the less abstract yet more vivid, more teeming life of the low end make the mind richer. All that is simple. What's harder to see is the whole world of mental attributes progressively transformed as the great sphere of mind turns on its axis.

The thinking world is fundamentally objective; the *being* world, subjective. You can think only in some *medium*—words or pictures or images, melody, harmony, numbers, physical gestures, pure ideas—which, *however* pure, the mind must somehow represent to itself. Thinking is the mental activity that needs a medium, and *any* medium is a potential method of communication. Thoughts can be expressed and therefore communicated. States of mental *being*, on the other hand, imply the existence of an attached body, because mental and physical states of being are intimately associated. States of being need not, *cannot*, be represented except by themselves. Green is a color state that cannot be represented except by itself; itchy is a physical state likewise; happiness is a mental state.

In a sense, then, thinking is an objective activity because its essence can be communicated. Some thoughts are easy to communicate, and some are nearly impossible. Yet we *have* represented that thought to ourselves, and we can represent it to other people too. A state of *being*, on the other hand, has no representation and cannot be communicated. It is strictly subjective. It can only be *experienced* by the person to whom it belongs and communicated indirectly, by pointing out which of the other person's repertoire of feelings equals your own current feeling.

There is, accordingly, yet one more way to understand the spectrum of consciousness: as a transition from what *can* to what *cannot* be said. Wittgenstein famously told us, in the very last sentence of the *Tractatus*, "What we cannot speak about we must pass over in silence."[1] He is wrong. At that moment, he was seeing only the spectrum's upper surface—in other words, not seeing the spectrum at all. Better to say, *What we cannot speak about we can still feel.* And even better: What we cannot speak about, we *must* still feel.

———

Although my earliest fascination with the mind was Freud-centered, I had no Freud bias when I began this project except a bias *against* the unthinking, uninformed rejection of Freud that I was seeing all over academia. Today, Freud is still mostly missing from academia. But as of this writing, troubled patients are again being referred (sometimes) to psychoanalysts—and are (sometimes) being helped. People are even reading Freud again. The pendulum swings; the whirligig of time brings in his revenges.

Still, Freud (like every genius) made many mistakes, failed to touch crucial topics, and above all, thought of the mind medically—as a sick organ to be healed. William Wordsworth took a broader view. Let us be "Prophets of Nature," he urged his old friend Coleridge, and show the world "how the mind of man becomes / A thousand times more beautiful than the earth / On which he dwells" (*Prelude*, Book 13). He even told us why the mind is beautiful: because the destiny it creates for us is "Effort, and expectation, and desire, / And something evermore about to be" (*Prelude*, Book 6).

Nothing is more beautiful than the human mind. Freud's is not the last word; it is—on human psychology proper, on depth psychology, in all of history—*the very first.* Let's return to depth psychology. It's time to move forward and learn new things.

Acknowledgments

My editors, Bob Weil and Phil Marino, were a crucial part of this project. They are professionals at the superb level that makes lawyers or financiers or physicists (not to mention ballplayers) rich and famous. They are maestros who rate with a Szell or von Karajan or Abbado. My copy editor, Stephanie Hiebert, went so far beyond the call of duty that she deserves a Distinguished Service Cross and an Eddy. (I trust that sleek gold Eddies will be handed out to the year's top editors at star-studded New York galas annually, starting soon.) My agents, Glen Hartley and Lynn Chu, made this happen through agenting genius.

I will always be grateful to Professor Drew McDermott of Yale for arguing with me about philosophy of mind and artificial intelligence endlessly over the years with amazing grace and patience—although, in his view, virtually everything I have ever said on the topic is wrong. Drew is thoughtful, he is brilliant, but (above all) he is the real thing. He shows the whole world how a scientist is supposed to operate. He is too modest; he is a hero in a field with far fewer heroes than it needs. David Berlinski and Jonathan Lear were generous with their time in ways that were essential to this book. But all three are *totally innocent* respecting the views expressed here. They endorse *nothing*!

An assortment of Yale students, especially the undergraduates of Computer Science 150 over the years, were essential to this project. The best of them—they know who they are—are as sharp, thoughtful, and intellectually aggressive as the best minds I have ever come across anywhere. They have never let me get away with a thing. My generation failed them—we made an ugly mess of their education—but they haven't failed us. We ought to be ashamed. As a nation we ought to put fixing our schools and colleges at the top of the list of urgent necessities instead of where we always do put it, in the sub-basement, or the trash.

I would be nowhere on this project or any other without four people who have always helped me think straight: Susan Arellano, Nick Carriero, Neal Kozodoy, and Martin Schultz. They don't know, they could never guess, how much they have done for me. They do it out of sheer generosity. All four rank among the mysteries of the universe.

Without my boys and without my wife, I wouldn't even be nowhere. My boys are grown up, and they are the best readers, critics, and fellow thinkers in the world. They and Jane keep me alive.

Literary Works Cited in the Text

Amis, Martin, *London Fields* (1989): 9, 119; *The Zone of Interest* (2014): 34

Austen, Henry, *Preface to Northanger Abbey and Persuasion* (1818): 71

Austen, Jane, *Emma* (1816): 23, 50, 67, 118; *Mansfield Park* (1814): 116, 117, 120; *Persuasion* (1818): 42, 70, 72, 117, 165; *Pride and Prejudice* (1813): 174

Bainville, Jacques, *Napoléon* (1931): 49

Banville, John, *Ancient Light* (2012): 35

Bellow, Saul, *Herzog* (1964): 115; *Humboldt's Gift* (1975): 60, 181; *Mr. Sammler's Planet*: 115

Blake, William, *Marriage of Heaven and Hell* (1793): 71, 234; *Songs of Innocence and of Experience* (1794): 234

Blixen, Karen, *Out of Africa* (1937): 35, 96

Brontë, Charlotte, *Jane Eyre* (1847): 44

Büchner, Georg, *Lenz* (1836): 34, 237

Chateaubriand, François-René de, *Mémoires d'outre-tombe* [*Memories from Beyond the Grave*] (1840): 176

Chekhov, Anton, "After the Theater" (1892; trans. by Robert Payne): 79

Churchill, Winston, *My Early Life* (1930): 116

Coetzee, J. M., *Age of Iron* (1990): 33; *Boyhood* (1997): 35, 218; *Childhood of Jesus* (2013): 59, 68; *Disgrace* (1999): 44; *Dusklands: The Narrative of Jacobus Coetzee* (1974): 35, 40; *Elizabeth Costello* (2003): 166; *In the Heart of the Country* (1977): 12, 118, 215; *Life & Times of Michael K* (1983): 92, 114; *The Master of Petersburg* (1994): 60, 82; *Slow Man* (2005): 118; *Summertime* (2009): 68, 72, 116, 216; *Youth* (2002): 89

Coleridge, Samuel Taylor, *Biographia Literaria* (1817): 26, 75; "Frost at Midnight" (1798): 228; "Kubla Khan" (1797): 237; *Remorse* (1813): 34

Donne, John, "Paradox VI" in *Paradoxes and Problems* (ca. 1600): 64; "The Second Anniversary" (1612): 71

Dostoyevsky, Fyodor, *The Brothers Karamazov* (1880; trans. by Richard Pevear and Larissa Volokhonsky): 194, 229; *Crime and Punishment* (1866; trans. by Constance Garnett): 68

Dumas, Alexandre, *Comte de Monte-Cristo* [*Count of Monte Cristo*] (1845): 91, 95

Eliot, T. S., *Dante* (Faber and Faber, 1930): 214; "Dante," in *Selected Essays* (Harcourt Brace, 1932): 3

Erpenbeck, Jenny, *The End of Days* (2012; trans. from the German by Susan Bernofsky): 122

Gide, André, *La porte étroite* [*The Narrow Gate*] (1909): 68, 165, 175

Glatstein, Jacob, *The Glatstein Chronicles* (1940; trans. from the Yiddish by Maier Deshell and Norbet Guterman): 82

Goethe, Johann Wolfgang von, *Faust*, Part I (1832): 216

Gordimer, Nadine, *The Pickup* (2001): 119

Hazlitt, William, *Characters of Shakespear's* [*sic*] *Plays* (1817): 237; "On the Feeling of Immortality in Youth" (1836): 139

Hemingway, Ernest, *Across the River and into the Trees* (1950): 67; *A Farewell to Arms* (1929): 33, 117; *The Sun Also Rises* (1926): 31, 69

Hölderlin, Friedrich, "Ehmals und jetzt" ["Then and Now"] (1799): 52

Hugo, Victor, *Notre-Dame de Paris* [*Hunchback of Notre Dame*] (1831): 91

James, Henry, *The Ambassadors* (1903): 117; *The Awkward Age* (1899): 113

Joyce, James, *Portrait of the Artist as a Young Man* (1916): 177, 185

Kafka, Franz, *The Castle* (1926; trans. by Edwin and Willa Muir): 238; "In the Penal Colony" (1919; trans. by Willa and Edwin Muir): 239; *The Metamorphosis* (1915; trans. by Will and Edwin Muir): 238; *The Trial* (1925; trans. by Willa and Edwin Muir): 238

Keats, John, "The Eve of St. Agnes" (1820): 51, 93, 177; "Ode to a Nightingale" (1819): 9, 51, 225; "Ode to Psyche" (1819): 51, 235

Mansfield, Katherine, "Bliss" (1918): 86

Meilhac, Henri, and Ludovic Halévy, *Carmen* (Libretto, 1875): 92, 93

Nabokov, Vladimir, *Ada* (1969): 42; *Speak, Memory* (1951): 26, 43, 76, 183

Naipaul, V. S., *Bend in the River* (1979): 181; *Enigma of Arrival* (1987): 12; *Half a Life* (2001): 24; *In a Free State* (1971): 221

O'Reilly, Sean, *Watermark* (2005): 59

Ozick, Cynthia, "Bloodshed" (1976): 119; *Foreign Bodies* (2010): 36, 50, 60, 116, 128, 148; *Heir to the Glimmering World* (2004): 87, 102; *Messiah of Stockholm* (1987): 61

Poe, Edgar Allan, *Marginalia* (1846): 236

Proust, Marcel, *À la recherche du temps perdu* [*In Search of Lost Time*] (1927): 23, 36, 50, 107, 200; "Contre Sainte-Beuve" ["Against Saint-Beuve"] (ca. 1900): 87

Pynchon, Thomas, *The Bleeding Edge* (2013): 119

Rémusat, Madame de, *Mémoires de Madame de Rémusat* (ca. 1821): 49

Rilke, Rainer Maria, *The Duino Elegies* (1923): 30, 164, 199

Rimbaud, Arthur, *Letters* (1871): 76

Robinson, Marilynne, *Absence of Mind* (2010): 17

Roth, Philip, *Deception* (1990): 142; *The Human Stain* (2000): 14, 34, 36; *The Humbling* (2009): 124; *Nemesis* (2010): 119; *The Prague Orgy* (1985): 70; *Sabbath's Theater* (1995): 34, 116

Salaman, Esther, *A Collection of Moments: A Study of Involuntary Memories* (1972): 34, 85

Shakespeare, William, *Coriolanus*: 6; *Hamlet*: 111, 120; *Julius Caesar*: 100; *King Lear*: 12, 16, 53, 59, 91, 119; *Love's Labours Lost*: 129; *Macbeth*: 90, 97, 158, 175, 237; *Othello*: 64, 88; *Richard III*: 102; *Romeo and Juliet*: 64; *The Tempest*: 31, 183; *Troilus and Cressida*: 19; *Twelfth Night*: 144; *The Winter's Tale*: 87

Shelley, Percy Bysshe, "Hymn of Apollo" (1824): 53; "Mont Blanc" (1816): 229

Svevo, Italo, *Zeno's Conscience* (1923; trans. by William Weaver): 40

Tolstoy, Leo, *Anna Karenina* (1877): trans. by Rosamund Bartlett: 90, 97, 175; trans. by Constance Garnett: 63, 145; *Death of Ivan Ilyich* (1886; trans. by Peter Carson): 120, 266; *War and Peace* (1869; trans. by Rosemary Edmonds): 118

Valéry, Paul, "Le cimitière marin" ["The Cemetery by the Sea"] (1920): 42

Wordsworth, William, "Ode: Intimations of Immortality" (1807): 225, 226; *Prelude*, Book 1 (1815): 22, 38; *Prelude*, Book 2 (1815): 93, 226; *Prelude*, Book 5 (1815): 4; *Prelude*, Book 6 (1815): 18, 251; *Prelude*, Book 13 (1815): 251; "Tintern Abbey" [full title "Lines Composed a Few Miles above Tintern Abbey"] (1798): 51, 118, 229, 230

Notes

Preface

1. John R. Searle, "Minds, Brains, and Programs," *Behavioral and Brain Sciences* 3 (1980): 417–24.

2. See especially Thomas Nagel, *Mind and Cosmos: Why the Materialist Neo-Darwinian Conception of Nature Is Almost Certainly False* (New York: Oxford University Press, 2012).

3. See, for example, Colin McGinn, *The Mysterious Flame: Conscious Minds in a Material World* (New York: Basic Books, 1999).

Chapter 1: The Tides of Mind

1. J. Allan Hobson, *Dreaming: An Introduction to the Science of Sleep* (New York: Oxford University Press, 2002), 122.

2. Stephen Grosz, *The Examined Life: How We Lose and Find Ourselves* (New York: Norton, 2013), 212.

3. John R. Searle, "Consciousness and the Philosophers," *New York Review of Books*, March 6, 1997.

4. John R. Searle, *The Rediscovery of the Mind* (Cambridge, MA: MIT Press, 1992), 21.

5. Jonathan Lear, *Love and Its Place in Nature: A Philosophical Interpretation of Freudian Psychoanalysis* (New York: Farrar, Straus, 2000), 3.

6. Ludwig Wittgenstein, *Philosophical Investigations*, trans. G. E. M. Anscombe (New York: Macmillan, 1953), II.iv, 178e.

7. İlham Dilman, *Philosophy as Criticism: Essays on Dennett, Searle, Foot, Davidson, Nozick* (New York: Continuum, 2011), 2.

8. Cited in Shaun Gallagher, *Phenomenology* (New York: Palgrave, 2012), 30.

9. Ibid., 29.

10. Lear, *Love and Its Place in Nature*, 11.

11. David J. Chalmers, *The Conscious Mind: In Search of a Fundamental Theory* (New York: Oxford University Press, 1996), 4.

12. Charles P. Siewert, *The Significance of Consciousness* (Princeton, NJ: Princeton University Press, 1998), 5.

13. Hobson, *Dreaming*, 15.

14. Ludwig Wittgenstein, *Culture and Value*, trans. Peter Winch (Chicago: University of Chicago Press, 1980), 19e.

Chapter 2: Three Thirds of the Spectrum

1. Ulric Neisser, "Literacy and Memory," in *Memory Observed: Remembering in Natural Contexts*, ed. Ulric Neisser (San Francisco: Freeman, 1982), 241.

2. David Foulkes, *Children's Dreaming and the Development of Consciousness* (Cambridge, MA: Harvard University Press, 1999), 145.

3. Stephen Grosz, *The Examined Life: How We Lose and Find Ourselves* (New York: Norton, 2013), 9.

4. Foulkes, *Children's Dreaming*, 16.

5. Cited in Jonathan Lear, *Freud* (New York: Routledge, 2005), 102.

6. Cited in Harold Bloom, *The Visionary Company: A Reading of English Romantic Poetry* (Garden City, NY: Doubleday, 1961), 143.

7. Inge Strauch and Barbara Meier, *In Search of Dreams: Results of Experimental Dream Research* (Albany: State University of New York Press, 1996), 234.

8. J. Allan Hobson, *Dreaming: An Introduction to the Science of Sleep* (New York: Oxford University Press, 2002), 5.

9. Ibid., 61.

10. David Foulkes, *Dreaming: A Cognitive-Psychological Analysis* (Hillsdale, NJ: Erlbaum, 1985), 12.

11. Hobson, *Dreaming*, 6.

12. Brian O'Shaughnessy, "The Id and the Thinking Process," in *Philosophical Essays on Freud*, eds. Richard Wollheim and James Hopkins (New York: Cambridge University Press, 1982), 107.

13. Cited in Lissa N. Weinstein, David G. Schwartz, and Arthur M. Arkin, "Qualitative Aspects of Sleep Mentation," in *The Mind in Sleep, Psychology and Psychophysiology*, 2nd ed., eds. Steven J. Ellman and John S. Antrobus (New York: Wiley, 1991), 197; italics mine.

14. Cited in George Dyson, *Turing's Cathedral: The Origins of the Digital Universe* (New York: Pantheon, 2012).

15. See William Lane Craig, *The Cosmological Argument from Plato to Leibniz* (London: MacMillan, 1980).

16. İlham Dilman, *Philosophy as Criticism: Essays on Dennett, Searle, Foot, David-son, Nozick* (New York: Continuum, 2011), 2.

Chapter 3: Every Day

1. Richard A. Griggs, *Psychology: A Concise Introduction*, 2nd ed. (New York: Worth, 2009), 68.

2. Jonathan Lear, *Love and Its Place in Nature: A Philosophical Interpretation of Freudian Psychoanalysis* (New York: Farrar, Straus, 2000), 11.

3. See Lear, *Love and Its Place*, 18, 26.

4. Cited in Jonathan Lear, *Freud* (New York: Routledge, 2005), 102.

5. Inge Strauch and Barbara Meier, *In Search of Dreams: Results of Experimental Dream Research* (Albany: State University of New York Press, 1996), 231.

6. Eric Klinger, *Daydreaming* (Los Angeles: Archer, 1990), 25.

7. See, however, Jerome Singer and Kenneth Pope's anthology *The Power of Human Imagination: New Methods in Psychotherapy* (New York: Plenum, 1978), and Donald Spence's *Narrative Truth and Historical Truth: Meaning and Interpretation in Psychoanalysis* (New York: Norton, 1982).

8. David Foulkes, *Children's Dreaming and the Development of Consciousness* (Cambridge, MA: Harvard University Press, 1999), 15; italics in the original.

9. See Dorthe Berntsen, "Involuntary Autobiographical Memories: Specula-tions, Findings and an Attempt to Integrate Them," in *Involuntary Memory*, ed. John H. Mace (Malden, MA: Blackwell, 2007), 24.

10. Jerome L. Singer, *Daydreaming and Fantasy* (New York: Oxford University Press, 1981), 108.

11. David Hume, *A Treatise of Human Nature* (New York: Oxford University Press, 1978).

12. Strauch and Meier, *In Search of Dreams*, 224.

13. Ibid., 218.

14. Cited in Leon Edel, *Stuff of Sleep and Dreams: Experiments in Literary Psychology* (New York: Harper & Row, 1982), 14.

15. Letter to Georges Izambard, May 13, 1871.

16. See, for example, Lear, *Freud*, 55ff; Josef Breuer and Sigmund Freud, "On the Psychical Mechanism of Hysterical Phenomena: Preliminary Communication," in *The Standard Edition of the Complete Psychological Works of Sigmund Freud*, vol. 2, trans. and ed. J. Strachey (London: Hogarth, 1955), 1–18. Note, on page 7, the famous observation *"Hysterics suffer mainly from reminiscences"* (italics in the original).

17. Cited in Ernest Jones, *The Life and Work of Sigmund Freud*, eds. L. Trilling and S. Marcus (Garden City, NY: Doubleday, 1963), 160.

18. Berntsen, "Involuntary Autobiographical Memories," 5.

19. For example, R. D. Ogilvie and J. R. Harsh, eds., *Sleep Onset, Normal and Abnormal Processes* (Washington, DC: American Psychological Association, 1994) is a collection of papers devoted entirely to sleep onset. The collection contains only a handful of transcripts.

20. Cited in David Foulkes, *Dreaming: A Cognitive-Psychological Analysis* (Hillsdale, NJ: Erlbaum, 1985), 13.

21. J. Allan Hobson, *Dreaming: An Introduction to the Science of Sleep* (New York: Oxford University Press, 2002), 41.

22. Stephen Grosz, *The Examined Life: How We Lose and Find Ourselves* (New York: Norton, 2013), 212ff.

23. David Gelernter, *Judaism: A Way of Being* (New Haven, CT: Yale University Press, 2009), 135ff.

Chapter 4: A Map

1. Jesse Prinz, "Emotion," in *Cambridge Handbook of Cognitive Science* (New York: Cambridge University Press, 2012), 193.

2. William James, *Psychology: Briefer Course*, in *Writings 1878–1899* (New York: Library of America, 1992), 350.

3. Jonathan Lear, *Love and Its Place in Nature: A Philosophical Interpretation of Freudian Psychoanalysis* (New York: Farrar, Straus, 2000), 122.

4. Ronald de Sousa, "The Mind's Bermuda Triangle: Philosophy of Emotions and Empirical Science," in *Oxford Handbook of Philosophy of Emotion*, ed. Peter Goldie (New York: Oxford University Press, 2010), 101.

5. Cited in Jonathan Lear, *Freud* (New York: Routledge, 2005), 23.

6. Goldie, *Oxford Handbook*, 4.

7. David Foulkes, *Dreaming: A Cognitive-Psychological Analysis* (Hillsdale, NJ: Erlbaum, 1985), 82. See also Marigold Linton, "Transformations of Memory in Everyday Life," in *Memory Observed: Remembering in Natural Contexts*, ed. Ulric Neisser (San Francisco: Freeman, 1982), 77–92.

8. Stephen Grosz, *The Examined Life: How We Lose and Find Ourselves* (New York: Norton, 2013), 89; italics mine.

9. Endel Tulving, *Elements of Episodic Memory* (Oxford: Clarendon, 1983).

10. Jerome L. Singer, *Daydreaming and Fantasy* (New York: Oxford University Press, 1981), 78.

Chapter 5: Spectrum, Upper Third: Abstraction

1. Ulric Neisser, "Memory: What Are the Important Questions?" in *Memory Observed: Remembering in Natural Contexts* (San Francisco: Freeman, 1982), 13.

2. Psychologists know that "the mind may, in fact, superimpose representations into composite memories." Janet Metcalfe, cited in *Relating Theory and Data: Essays on Human Memory in Honor of Bennet B. Murdock*, eds. W. E. Hockley and S. Lewandowsky (Hillsdale, NJ: Erlbaum: 1991). The *point* of such superimposition doesn't clearly emerge in the literature. But the psychologist Janet Metcalfe cites the mind's ability to superimpose in a fascinating context: She discusses a late-nineteenth-century fad for superimposing photographs of individuals of some particular variety in order to see, supposedly, a typical specimen. Photos of "twelve mathematicians" or "sixteen naturalists" were said, when superimposed, to yield images of a typical mathematician or typical naturalist. In forming its focused abstrac-

tions, the mind does *not* literally superimpose recollected images. But its operations have that effect.

3. I proposed learning-by-forgetting and abstraction as a natural control on the growth of memory, as mental realities and software techniques long ago (*The Muse in the Machine: Computerizing the Poetry of Human Thought* [New York: Free Press, 1994]). They are still obvious and powerful.

4. Georges Rey, *Contemporary Philosophy of Mind: A Contentiously Classical Approach* (New York: Blackwell, 1997), 181.

5. Here's an index of how natural the idea is: Stephen Grosz reports a patient's dream in which she is riding on a train. She herself interprets it for him: the train stands for her *train of thought*. The dream handles the metaphor matter-of-factly, like the normal part of everyday language that it is.

6. Will your thinking computer be capable of this simple maneuver? A quick check on an invented image? There's no reason a computer shouldn't be able to do this sort of thing. But the details are as tricky as they are critical to a well-working imitation mind.

Chapter 6: Spectrum, Middle Third: Creativity

1. Roger Penrose, *The Emperor's New Mind: Concerning Computation, Mind and the Laws of Physics* (New York: Oxford University Press, 1989), 429.

2. George Steiner, *Real Presences: Is There Anything in What We Say?* (Boston: Faber & Faber, 1989), 181.

3. J. Metcalfe and D. Weibe, "Intuition in Insight and Noninsight," *Memory and Cognition* 15 (1987): 238–46.

4. E. R. Dodds, *The Greeks and the Irrational* (Berkeley: University of California Press, 1951), 11.

5. J. A. Fodor, *The Modularity of Mind: An Essay on Faculty Psychology* (Cambridge, MA: MIT Press, 1983), 107.

6. We could, in principle, announce a resemblance in view of the similar aspect ratios of the two faces, or the exact same cubic inches of nose. But we probably wouldn't.

7. Philosophers and psychologists have worked hard for decades—as I have discussed—trying to explain how analogies are put together, how creativity works. Often their explanations wrestle with two subproblems: "putting up"

and "shooting down." "Putting up" means proposing possible creative solutions, or new and original analogies. "Shooting down" means getting rid of bad or invalid possibilities.

Some thinkers call the process itself a solution: to be creative means that you're good at generating large numbers of possibilities and rejecting the bad ones.

One reason this answer is unsatisfying is its failure to match the accepted psychological profile of finding a creative solution. *Eureka*—"I have found it" in Archimedes's Greek—is a famous exclamation partly because it exactly matches everyday experience. There are few creative geniuses among us, but nearly everyone hits on a creative idea occasionally. We know what a eureka moment feels like. When it happens, we tend to *recognize* it as we do a friend's face. We *feel* that we have discovered a solution or a valid new analogy, that we have accomplished something—that inspiration has arrived out of the blue. We have all experienced those moments of recognition. But if, on the other hand, you have found the right answer by beating the bushes exhaustively, searching through a pile of wrong ideas, and finally finding a solution by sheer persistence, you are unlikely to feel *inspired*—unlikely to feel that the idea came to you suddenly, out of the blue.

Some thinkers have resorted to quantum mechanics to explain how a large collection of possibilities could suddenly yield one correct answer. It's an interesting, exotic, wholly unnecessary maneuver. Mind scientists don't pay enough attention to emotion and sensation, feeling and simple *being*. The answer is right under their noses—but not in the upper spectrum, which is the only place they like to spend time.

8. John Carey, *John Donne, Life, Mind and Art* (New York: Oxford University Press, 1981), 138.

9. Cited in A. Paivio, "Psychological Processes in the Comprehension of Metaphor," in *Metaphor and Thought*, ed. A. Ortony (New York: Cambridge University Press, 1979), 159.

10. Cited in B. Willey, *Samuel Taylor Coleridge* (New York: Norton, 1957), 96.

11. Morton F. Reiser, *Memory in Mind and Brain: What Dream Imagery Reveals* (New York: Basic Books, 1990), 11.

12. Ibid., 46.

13. Ibid., 47.

14. Ibid., 92.

Chapter 7: Spectrum, Lower Third: Descent into Lost Time

1. Eric Klinger, *Daydreaming* (Los Angeles: Archer, 1990), 7; italics mine.
2. Jerome L. Singer, *Daydreaming and Fantasy* (New York: Oxford University Press, 1981), 78.
3. David Foulkes, *Dreaming: A Cognitive-Psychological Analysis* (Hillsdale, NJ: Erlbaum, 1985), 18.
4. Gerald W. Vogel, "Sleep-Onset Mentation," in *The Mind in Sleep: Psychology and Psychophysiology*, 2nd ed., eds. Steven J. Ellman and John S. Antrobus (New York: Wiley, 1991), 126.
5. Victor Terras, *A Karamazov Companion* (Madison: University Wisconsin Press, 1981), 103.
6. Vogel, "Sleep-Onset Mentation," 128.
7. Ibid.
8. J. Allan Hobson, *Dreaming: An Introduction to the Science of Sleep* (New York: Oxford University Press, 2002), 113.
9. "Those about him did not understand or would not understand it, but thought everything in the world was going on as usual. That tormented Ivan Ilyich more than anything" (Tolstoy, *The Death of Ivan Ilyich*).
10. Sigmund Freud, *The Interpretation of Dreams II: Standard Edition of the Complete Psychological Works of Sigmund Freud*, vol. 5, trans. and ed. V. J. Strachey (London: Hogarth, 1955), 571; italics in the original.

Chapter 8: Where It All Leads

1. Vivian Gussin Paley, *Wally's Stories* (Cambridge, MA: Harvard University Press, 1981), 18.
2. James Garbarino, Frances M. Stott, and the Faculty of the Erikson Institute, *What Children Can Tell Us: Eliciting, Interpreting and Evaluating Critical Information from Children* (San Francisco: Jossey-Bass, 1992), 78.
3. R. M. Billow, "Observing Spontaneous Metaphor in Children," in *Child Language: A Reader*, eds. M. B. Franklin and S. S. Barten (New York: Oxford University Press, 1988).
4. Gerald W. Vogel, "Sleep-Onset Mentation," in *The Mind in Sleep, Psychology and Psychophysiology*, 2nd ed., eds. Steven J. Ellman and John S. Antrobus (New York: Wiley, 1991), 131.

5. John H. Flavell, *Cognitive Development* (Englewood Cliffs, NJ: Prentice-Hall, 1977), 65.

6. Paley, *Wally's Stories*, 1.

7. Garbarino et al., *What Children Can Tell Us*, 68.

8. Paley, *Wally's Stories*, 8.

9. Ibid., 29.

10. Garbarino et al., *What Children Can Tell Us*, 47.

11. Flavell, *Cognitive Development*, 75.

12. Ibid.

13. This was brought home to me years ago in a gigantic department store in Manhattan by my future wife. On a Sunday afternoon, we had trekked and shopped our way from one vast windowless floor to the next along a convoluted path of clusters, zones, counters, racks, and display shelves: spirals within spirals. At last we arrived at a zone that had a pleated linen skirt, not too long, in which she looked lovely. That concluded our shopping. (I hoped forever.) Time to get out. How? We had corkscrewed around for so long, I had no idea where we stood relative to the actual world. We were in the middle of a vast waste of noise and crowds and women's fashion; for all I knew, we might have been down a coal mine in West Virginia. "Shouldn't we head towards Thirty-Fourth Street?" she said calmly, and pointed. We walked off in that direction. Right!

14. Dorothy G. Singer and Jerome S. Singer, *The House of Make-Believe: Children's Play and the Developing Imagination* (Cambridge, MA: Harvard University Press, 1990).

15. Peter Ackroyd, *Blake* (New York: Knopf, 1996), 30.

16. Ibid., 35.

17. Ibid., 34.

18. David Foulkes, *Children's Dreaming and the Development of Consciousness* (Cambridge, MA: Harvard University Press, 1999), 6ff.

19. Ruth Morse, "Daggers Drawn," *Times Literary Supplement*, July 3, 2015.

Chapter 9: Conclusions

1. Ludwig Wittgenstein, *Tractatus Logico-Philosophicus*, trans. D. Pears and B. McGuinness (London: Routledge & Kegan Paul, 1922).

Index

Page numbers in *italics* refer to illustrations.

ABOUT THE AUTHOR

David Gelernter is a BA (Yale 1976), MA in classical Hebrew (Yale 1977), and PhD in computer science (SUNY at Stony Brook, 1983). He is now a professor of computer science at Yale. His work in the 1980s with Nick Carriero on the parallel programming system Linda led to the development of high-speed database search techniques that subsequently proved important to several leading Web-search efforts and companies. His 1991 book, *Mirror Worlds: Or, The Day Software Puts the Universe in a Shoebox . . . How It Will Happen and What It Will Mean*, "foresaw" the World Wide Web (Reuters) and was called "one of the most influential books in computer science" (*Technology Review*, July 2007). It led to the development of the Web programming language Java. His 1990s work with Eric Freeman on the Lifestreams system created the first modern social network (see *Wall Street Journal*, October 2, 2015) and predicted the rise of blogs, activity streams, Twitter streams, and other time-ordered data feeds. In recent years he has published pieces about technology and society (mostly in *Commentary*, the *Wall Street Journal*, and the *Frankfurter Allgemeine Zeitung*), worked on the "Glassbooks"© alternative to conventional ebooks, and—mainly—pursued research that has led to *The Tides of Mind*.

Gelernter has also served for several years on the advisory board of the National Endowment for the Arts (appointed by President George W. Bush) and has shown his paintings many times; his first museum show was *Shm'a/Listen: The Art of David Gelernter* (2012) at the Yeshiva University Museum in lower Manhattan. His works are in the permanent collections of the Yeshiva University Museum, the Tikvah Foundation, and several distinguished private collectors.